PSPICE FOR BASIC CIRCUIT ANALYSIS

JOSEPH G. TRONT

Virginia Tech

Higher Education

Boston   Burr Ridge, IL   Dubuque, IA   Madison, WI   New York   San Francisco   St. Louis
Bangkok   Bogotá   Caracas   Kuala Lumpur   Lisbon   London   Madrid   Mexico City
Milan   Montreal   New Delhi   Santiago   Seoul   Singapore   Sydney   Taipei   Toronto

Higher Education

PSPICE FOR BASIC CIRCUIT ANALYSIS

1 2 3 4 5 6 7 8 9 0 DOC/DOC 0 9 8 7 6 5 4

ISBN 0–07–293981–8

Publisher: *Elizabeth A. Jones*
Senior sponsoring editor: *Carlise Paulson*
Developmental editor: *Michelle L. Flomenhoft*
Marketing manager: *Dawn R. Bercier*
Senior project manager: *Jane Mohr*
Senior production supervisor: *Laura Fuller*
Cover designer: *Rick D. Noel*
Cover image: *©PhotoDisc, #bs28028m,* schematic diagram
Compositor: *Schiesl Outside Services*
Typeface: *11/13 Times Roman*
Printer: *R. R. Donnelley Crawfordsville, IN*

Library of Congress Cataloging-in-Publication Data

Tront, Joseph G.
    PSpice for basic circuit analysis / Joseph G. Tront. — 1st ed.
        p.   cm.
    Includes index.
    ISBN 0–07–293981–8
    1. Electric circuit analysis—Data processing. 2. Electronic circuit design—Computer-aided
        design. 3. PSpice. I. Title.

TK454.T764   2004
621.319'2'0113—dc22   2003062407
                                                         CIP

Contents

Preface

PSpice for Basic Circuit Analysis introduces students to the fundamental uses of PSpice in support of basic circuit analysis. PSpice is a very powerful circuit simulation program that can assist a circuit designer in the process of solving multi-element circuits in a small amount of time. This book may be readily used to support basic circuit analysis courses and the textbook used for those courses. It contains detailed explanations and examples of the use of PSpice in typical problem solving situations. The book is designed so that the student may advance rapidly to solving circuit analyses problems typical of those contained in undergraduate electrical and computer engineering courses. Although we only use the very fundamental capabilities of PSpice, the principles can be easily extended to analyze the complex electrical and electronic networks used in modern integrated circuit design.

One of the best ways to promote the appropriate use of computer-aided design tools like PSpice is to integrate their use directly into the basic fabric of the course. To encourage this integration, the topics and chapters of this book follow the typical ordering of many of the popular textbooks used in circuit analysis classes. Instructors will find it beneficial to encourage students to perform a mixture of analyses including solving a rich set of "by-hand" problems as well as a set of problems that are solved using PSpice.

Operational features of OrCAD Capture and the plotting program Probe are described in this book. Students can use this material to become proficient with the entry of circuit descriptions and the output of simulation results. We also discuss a method of entering circuit information in netlist format so that students can get a minimal understanding of one of the basic operations of PSpice.

Examples shown throughout the book are based on the OrCAD software package PSpice version 9.2. However, the principles established in the book are applicable to many of the other versions of SPICE that are widely available both academically and commercially. A copy of the OrCAD PSpice 9.2 Demo CD is packaged with this book. Updates, newer demo versions, as well as commercial versions can be obtained at www.cadence.com/products/pspice/.

Installing PSpice is fairly simple. If you have the PSpice disk in hand, place it in the CD-ROM drive and the automated installation process will guide you through the process. If the autorun program does not start up within a few seconds after you insert the CD, click on the Start button in the bottom left corner of your screen. Click on Run.... Type in the following command:

d:\setup   where d: is the logical name of your CD-ROM drive. If you have downloaded demo PSpice software from the Cadence web site or from other distribution sites, follow the instructions for installation found on the web site. Generally, the software can be installed by double-clicking on the downloaded filename in Windows Explorer.

The author would like to thank Cadence Design Systems, Inc. for allowing McGraw-Hill to distribute PSpice Release 9.2 Educational Software with this book.

Chapter 1 Introduction

1.1 Background

SPICE is a very powerful general purpose analog circuit simulator that is used to validate circuit designs and to predict circuit behavior. SPICE is an acronym that stands for **S**imulation **P**rogram with **I**ntegrated **C**ircuit **E**mphasis. PSpice is a derivative of the original SPICE program that includes several added values such as a graphical user interface and an enhanced plotting capability, which make it into a viable commercial product. The original focus of SPICE was to support integrated circuit design, though now PSpice and other derivatives find a much broader use in supporting general circuit design.

SPICE was originally developed at the University of California at Berkeley in 1975 and was described in a thesis entitled "SPICE: A Computer Program to Simulate Semiconductor Circuits" written by L. Nagel. The original program was subsequently further described in a number of papers written by L. Nagel, D. Pederson, E. Cohen, A. Vladimirescu and S. Liu.

PSpice can perform several different types of circuit analyses. Amongst those most important are:

- DC analysis: calculates the DC transfer curve
- Transient analysis: calculates outputs as a function of time when a large signal input is applied
- AC Analysis: calculates the outputs as a function of frequency
- Fourier analysis: calculates and plots the frequency spectrum of the response
- Noise analysis
- Sensitivity analysis
- Distortion analysis
- Monte Carlo Analysis

All analyses can be done at different temperatures. The default temperature is 300° K.

To accomplish all of the different types of circuit analyses, a simulator must contain mathematical models of the various circuit components of interest. PSpice contains internal models for the following components:

- Independent and dependent voltage and current sources
- Resistors
- Capacitors
- Inductors
- Mutual inductors
- Transmission lines
- Operational amplifiers
- Switches
- Diodes
- Bipolar transistors
- MOS transistors
- JFET
- MESFET

PSpice also has analog and digital libraries containing models for other components such as NAND and NOR gates, flip-flops, and other digital gates, op amps, etc. This makes it a useful tool for a wide range of analog and digital applications.

1.2   The Design Process

Circuit analysis is one of the elementary steps in performing circuit design. Designing a circuit is generally an iterative process that typically starts by setting up performance requirement specifications. The designer next produces a trial circuit targeted at satisfying the specifications. An analysis is then performed to determine if the current rendition of the circuit satisfies the original design constraints. If the specifications are met, the design process is complete. If not, the cycle begins again starting with the synthesis of a new/modified circuit and proceeding to the analysis and specification satisfaction-checking steps until the overall design is correct. Occasionally the designer must go back and re-examine the original specifications based on a better understanding of the desirable behavior obtained during the design process. Overall, the design process can be described by the flow model shown in Figure 1. Circuit analysis tools like PSpice can speed up the design process by performing the repeated analysis task. Results obtained from a PSpice analysis are used by the designer to decide on how the circuit should be modified to more closely meet the specs. Once a circuit configuration has been set up and run through PSpice, it is relatively easy to modify component values and sub circuit configurations as the designer iterates toward a final solution. However, the user must take care to use PSpice wisely and base new circuit iteration on not only the previous output of the PSpice analysis, but also on the designer's understanding of the operation of the electric circuit behavior. In other words, PSpice is not the magic orb that will solve all of your circuit design problems—you must have a sound understanding of electric circuit analysis to use PSpice effectively.

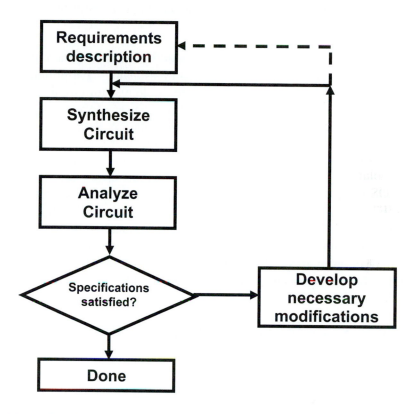

Figure 1 Flowchart for the typical circuit design cycle.

1.3   Appropriate Use of CAD

Computer-aided design software has become an integral part of the engineer's toolkit. CAD software packages can dramatically speed up the design and implementation cycle by assisting the engineer in organizing their ideas, analyzing design performance, testing systems in simulated environments, and with some advanced tools, suggesting modifications that bring the design closer to the desired specifications. With all of this power, there are still potential pitfalls for careless users of CAD tools. These pitfalls can be avoided through the use of common sense and an awareness of the fact that CAD tools are not infallible.

The user must first understand that even an excellent CAD tool is no substitute for a good fundamental understanding of the related subject matter. In the case of PSpice, the user must be able to analyze basic circuits to know whether or not the result presented by PSpice is within the expected range of solutions. Users should perform a *reasonableness check* of every simulation solution set to know whether or not PSpice is producing a result that is fully viable for the circuit under analysis.

Gross reasonableness checking can be performed by examining the magnitudes of the data points in the solution to determine if the circuit could reasonably have produced the voltage or currents being reported. For example, if you have a multiple resistor circuit with a few different voltage supplies, none of which is larger than 10 volts, the user can reasonably expect that none of the voltage in the circuit would be higher that 10 volts. A data point of 50 volts for this circuit should cause the user to closely examine the overall results. Circuit connections should be checked, element models should be scrutinized, and other components of the overall simulation should be reviewed to insure that any specification errors are eliminated. Another simple reasonableness check involves studying the end points of a dataset and the points of critical inflexions. For example, if you plot the voltage across an element as a function of time, you should perform rough calculations of the expected voltage at the initial time and the final time in the plot. You should also examine the output when plotted results contain sharp transitions or have discontinuities in areas where none might be expected. These types of precautions will help keep the user from inadvertently using erroneous results.

1.4   *Versions of SPICE and Limitations*

As mentioned earlier, PSpice is derived from the program called SPICE, which in fact, follows upon the design of earlier circuit design programs. The list of SPICE predecessors includes ECAP, SCEPTRE, PCAP, NCAP, CircuitPro, Super SCEPTRE, CANCER, and others. Further development of SPICE produced the widely used SPICE2 and SPICE3. Several commercially available versions of SPICE have been developed based on the original work. Besides PSpice, programs such as HPSICE still enjoy broad use. Commercializing the software involved re-writing some of the code to make it more reliable and more supportable. Enhancements in speed and convergence are important improvements produced through commercialization. Increased numbers of built-in models and larger external model libraries are also products of marketing and competitive pressures. Some companies have developed their own proprietary circuit simulation packages, several of them modeled after SPICE. Through all of this development, the root software SPICE and its various derivatives have become the de facto standard for computer-aided circuit analysis tools and have grown to be used by a wide variety of circuit designers.

Designers must always recognize the limitations of their tools and use them wisely. First and foremost, the user must understand that PSpice does not perform design, but is only a tool to perform analysis. PSpice is useful during the design cycle, but only when the user understands where it fits in the process. Secondly, the result that PSpice provides is a numerical solution as opposed to an analytical solution. It may be necessary for a designer to perform several analyses in order to understand in general how design modification affects the circuit behavior. This understanding is necessary in order to perform circuit synthesis efficiently. Analytical results are obtained by applying the principles learned in a circuit analysis class. Combining good analytical capabilities with the ability to rapidly produce numerical results leads to a very effective design process.   Finally, there are some circuits that PSpice is incapable of solving due to lack of numerical convergence or other problems. In many cases, the circuit model can be modified into a different but equivalent form that produces satisfactory results. Of course, the model translation must be done carefully so that the new model accurately represents the original circuit.

The commercial version of PSpice is capable of analyzing circuits with a very large number of nodes and elements depending on which version you purchase. PSpice demo version 9.2 is limited to:

- 64 nodes
- 10 transistors
- 65 digital primitive devices
- 10 transmission lines in total (ideal or non-ideal)
- 4 pair-wise coupled transmission lines.

There are additional limitations on the contents of the model libraries, on the number of parts that can be in a schematic drawing, the user's ability to edit device characterization parameters, and other more advanced capabilities. Even with these limitations, PSpice is a more than capable tool for basic circuit analysis courses.

Chapter 2 Getting Started

2.1   Circuit Description

The first step in the process of simulating the behavior of a circuit is to describe the components and connections of the circuit in a form that the PSpice software can understand. PSpice uses a file called a *netlist* as the input to be analyzed and simulated. A netlist consists of a set of single text lines that contain information about the component type, the nodes to which the component is connected, the size of the component, and possibly other relevant component values. Netlists can be entered by hand, however, the OrCAD version of PSpice has a schematic capture program associated with it that allows the user to enter circuit descriptions in graphics form and have them subsequently converted to a netlist for analysis with PSpice. We will first show how a netlist may be generated using OrCAD Capture after which we will examine the process of generating a netlist by hand.

Example 1: Use the OrCAD Capture program to produce a netlist for the simple resistive circuit shown in Figure 2.

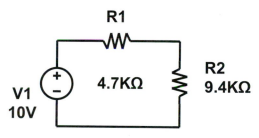

Figure 2 Simple resistive circuit.

Solution 1: Begin by starting the OrCAD Capture program by clicking on the Start button, selecting All Programs, PSpice Student, Capture Student (symbolized in this book by Start>All Programs>PSpice Student>Capture Student). Click on File>New>Project in the OrCAD Capture window which will bring up the display shown in Figure 3.

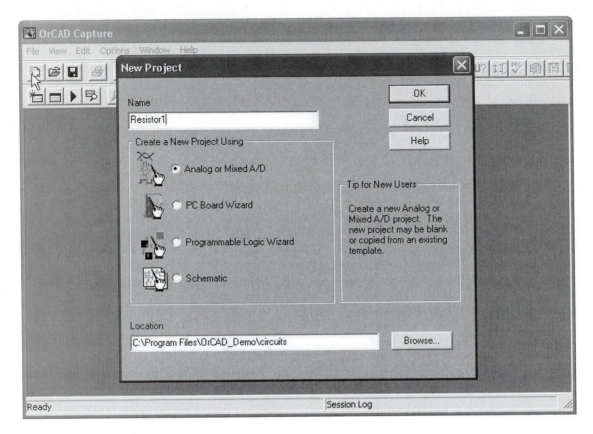

Figure 3 OrCAD Capture window.

Type in a name for the project in the area labeled Name. This name will be used to identify the project in subsequent simulation windows and data files that are generated by OrCAD or PSpice. Next, click on the radio button next to Analog or Mixed A/D causing the OrCAD Capture program to prepare a simulation file that can be used by PSpice. We use the name Resistor1 for this example. Finally, we specify the location where we want the files associated with this simulation project to be placed. We may either type in the full path specification starting with the drive letter specification through all subdirectory names, or we may use the Browse feature of Windows to specify the location. Once all of the parameters in this window have been filled in, we click on the OK button. We have specified a location in the subdirectory C:\Program Files\OrCAD_Demo\ciruits after having created it inside the C:\Program Files\OrCAD_Demo directory using the button labeled Create Dir… from the Select Directory Window.

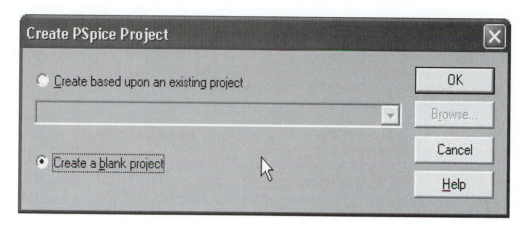

Figure 4 OrCAD Capture dialog used to select a project template.

The Capture software will then bring up the window shown in Figure 4 asking us to choose a template around which this capture process will function. Click on the radio button labeled Create a Blank Project and complete the selection by choosing OK. The next screen that appears is shown in Figure 5 and contains the window in which we enter the schematic diagram along with a window that contains information about the files that are part of this capture project. The schematic window contains a drawing grid with parts set by default to snap to the grid.

To begin entering circuit components on the schematic, click on the button, second from the top on the right-hand side, which contains the icon that looks like ⊲. A window similar to the one in Figure 6 will appear asking us to choose the part that we want to place in the drawing area. Depending on the installation of OrCAD and whether or not anyone else has already used the software before, there may or may not be a string of part names in the Part List area or a set of parts library names in the Libraries area. Libraries may be added to the list by clicking on the button labeled Add Library and choosing any libraries appropriate to the design. Standard libraries are located in the subdirectory c:\Program Files\OrCAD_Demo\Capture\Library\Pspice and have the file type .olb. We will use parts from the analog.olb and source.olb library files, so be sure to add these libraries at the beginning. Click on ANALOG under the Libraries area. A list of parts contained in the analog library will appear. Click on the part labeled R in the Part List. A symbol for a resistor will appear in the lower right side of the window. The default value for the resistor is set to 1 kΩ, while the default name is set to R?, where the question mark is replaced by

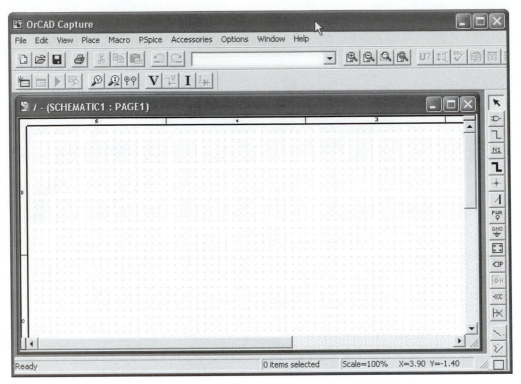

Figure 5 Schematic entry window.

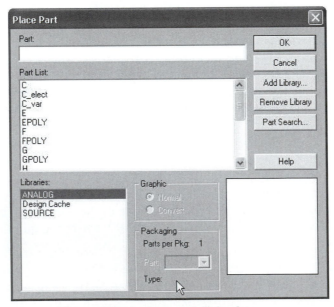

Figure 6 Parts placement dialog box.

a sequence number each time an instance of the resistor is placed in the schematic window. Numbering starts from 1 to n and may be changed after a resistor is placed in the schematic.

Click on the OK button in the dialog box. The dialog disappears and the cursor has the diagram of a resistor tracking along its movement path. By clicking on the left mouse button you will drop an instance of the resistor onto the grid of the schematic window.

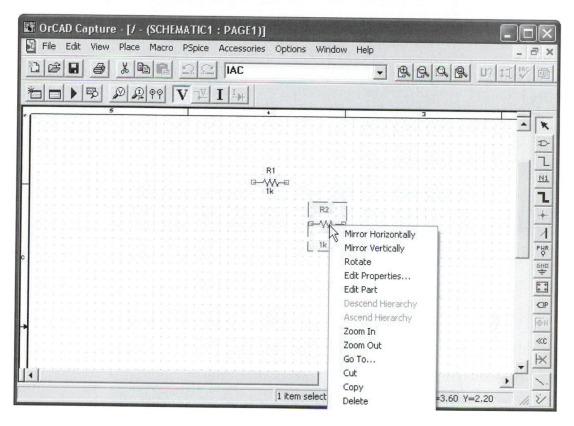

Figure 7 Right-click to obtain an individual part manipulation menu.

Place a resistor in a position just above the center of the schematic window. Next, place another resistor lower and to the right of the first resistor. Hit the Esc key to end the parts placement operation. With the last resistor still selected following its placement, click Ctrl-R on the keyboard. This causes the selected part, the second resistor in this case, to rotate 90° in the schematic. A part may also be similarly manipulated by right-clicking on a part that is selected and then bringing up a menu of operations as shown in Figure 7. Rotate the R2 resistor so that it is placed vertically and then double-click on the label indicating the default value of 1k for the resistor. A Display Properties dialog box like the one shown in Figure 8 will appear. (We may have to click Esc to de-select the entire resistor part and then click on just the characters "1k" in order to bring up the properties for the resistor value). Change the resistor size in the area labeled Value to be 9.4K as in the example of Figure 2. We may also change the name of the resistor by double-clicking on the R2 next to the resistor part. A Display Properties dialog box similar to the one in Figure 8 will be shown, the difference being that this dialog allows us to change the Part Reference, which is the name displayed on the schematic. Other parts such as voltage sources, capacitors, inductors, etc., will have adjustable properties similar to those shown in Figure 8.

When changing a Part Reference, any valid printing character may be placed in the Value field. To change the size value assigned to a part, numbers placed in the Value field must either be integers or real numbers. Integers can be either positive or negative, e.g., 14, or -642. Real numbers can be numbers containing decimal points, (e.g., 9.807) or numbers containing integer exponents (e.g., 1.602E-19), or numbers containing a symbolic exponent (e.g., 0.2998G). In the latter example, the G symbolizes an exponent of 10^9. A list of acceptable symbolic scale factors for PSpice is shown in Table 1.

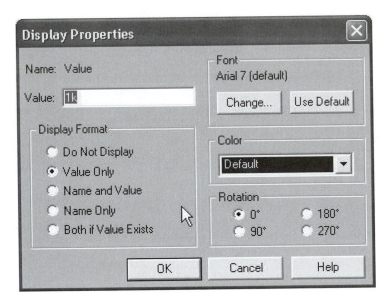

Figure 8 Default Display Properties dialog for a resistor.

Table 1 PSpice Scale Factors

Symbolic Suffix*	Mnemonic	Exponential Form	Value
F	femto	1E-15	10^{-15}
P	pico	1E-12	10^{-12}
N	nano	1E-9	10^{-9}
U	micro	1E-6	10^{-6}
M	milli	1E-3	10^{-3}
K	Kilo	1E3	10^{3}
MEG	mega	1E6	10^{6}
G	giga	1E9	10^{9}
T	tera	1E12	10^{12}

* Note that in PSpice, a suffix may be either a capital or lower-case letter.

Change the value of the resistor R1 to be 4.7 KΩ matching the value in Figure 2. Now let's place the DC power source in the schematic. Click on the place parts icon ⊡ and choose the SOURCE library in the lower left area. Then double-click on VDC in the Part List area and place the DC voltage source part in the left part of the schematic. Change the value of the voltage source to be 10, in the same manner that you changed the values of the resistors.

Now that we have selected all of the parts, placed them in the schematic, and set their Values and Reference Names, we are ready to wire the components together to form a circuit. From the buttons menu on the right side of the Capture screen, click on the Place Wire button ⌐ and the drawing cursor will change to a cross-hair and allow us to stretch wires between component nodes. Start by clicking the cross-hair on the top node of the power source symbol. Move the cross-hair to the left node of the resistor symbol and click again. Capture will stretch a wire between the two nodes, squaring off a 90° angle as the wire turns corners. Wire all three parts together so that the schematic looks something like the one in Figure 9.

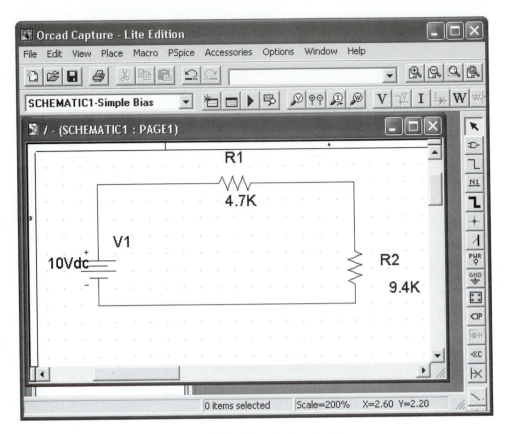

Figure 9 Simple resistor circuit.

The wire drawing capability is quite flexible, allowing users to connect wires to component nodes or to join wires to other wires. Wires may be ended at a location other than a component connection node by clicking at the desired end point and hitting the Esc key. Wires can then be joined end node to end node. Segments of wires may also be selected and stretched, moved, or deleted in the manner of a typical CAD drawing. When placing wires, we need to be careful that we do not have unexpected disconnects. Save the schematic, then try placing, stretching, and deleting a few wires around the schematic. Afterward, get the schematic back to the form of the one shown in Figure 9.

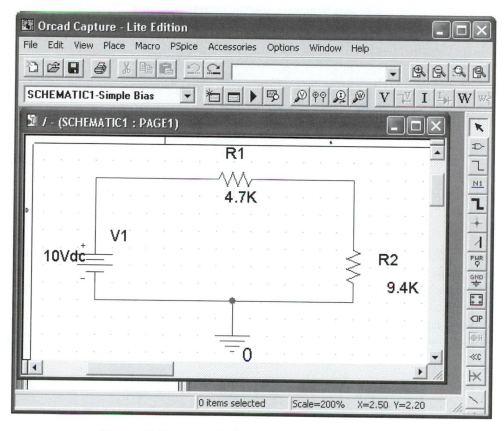

Figure 10 Simple resistor circuit with PSpice ground.

In order for PSpice to be able to work with the circuit, we must place a special reference node called GND and referred to as symbol **0** in OrCAD Capture. Click on the GND button 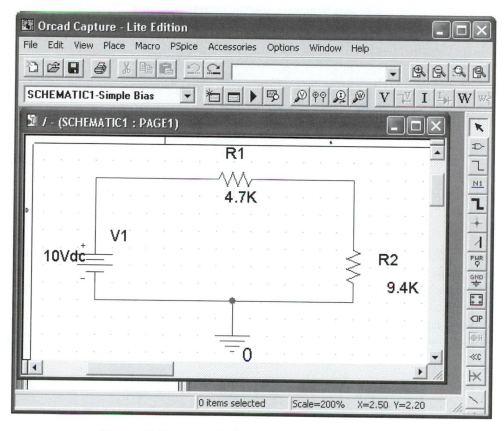 in the button menu on the right of the Capture screen. A Place Ground dialog box will appear. In the Libraries area on the left bottom of the dialog box, select the SOURCE library. (If the SOURCE library does not appear in this area, click on Add Library and select the file source.olb from the ..OrCAD_Demo>Capture>Library>PSpice subdirectory). Choose the symbol **0** as shown in Figure 10. Place the ground symbol in the schematic and connect it with a wire as shown in Figure 11. The symbol **0** is a special symbol that produces what is called the "reference node" for PSpice. Other ground symbols found within Capture will cause failures in the PSpice simulation and should be avoided. Be sure to remember to insert the PSpice ground symbol into the circuit before simulation, for without it the circuit will not simulate.

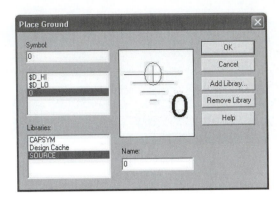

Figure 11 Place Ground dialog box.

We have now performed all of the steps necessary to describe a simple circuit in preparation for PSpice analysis. The process we just completed is generally referred to as schematic capture. The next automated operation that will be performed is a translation of the circuit information provided in the schematic, into a machine readable description called a netlist. Later in this chapter we will analyze the netlist generated by the Capture program. We will then examine how you may directly generate the netlist and simulate the circuit without using the capture program.

Before a simulation can be executed, we must choose the type of analysis that is to be performed by PSpice. The next section shows us how to tell PSpice the type of analysis to perform, and the type of results you are interested in seeing.

2.2   *Specifying the Analysis*

PSpice can perform several different types of analyses on the schematic we have drawn. For a simple resistive circuit like the one in Figure 11, the only analyses that are meaningful are analyses of the dc behavior of the circuit. PSpice will perform two different types of dc analysis: a bias point analysis and a .DC analysis. In the bias point analysis, PSpice analyzes the circuit's response to the basic dc sources operating in conjunction with the resistive elements in the circuit. The bias point analysis is also called the operating point analysis and sometimes symbolized as the .OP analysis.

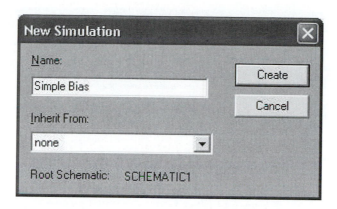

Figure 12 Dialog box to establish a simulation name.

In order to differentiate between the simulations performed, the first step is to name the simulation. Naming is done by choosing the menu option PSpice>New Simulation Profile. This selection generates the dialog box shown in Figure 12 in which we enter a name for this particular simulation run. We have named the simulation for Example 1, Simple Bias. Once we click on the Create button, a secondary dialog box appears in which we are asked to choose the analysis type for this simulation. In Figure 13 we see a Simulation Settings dialog box in which we have chosen the Analysis Type to be Bias Point. Several other settings are available in this dialog and its associated tabs. For now, leave them fixed to the default settings. One thing we may want to note for future reference is the location where the simulation output files will be placed. In the Simulation Settings dialog box under the general tab, we can see the area marked Output, and for this simulation we will note that the output files will be located at ...\OrCAD_Demo\circuits\resistor1-SCHEMATIC1-Simple Bias.out. We are now ready to initiate the PSpice simulation once we click OK to close this dialog box.

Simulation begins once we click on the PSpice>Run menu item in the Capture window. A display window similar to the one shown in Figure 14 will appear showing the initial results of the simulation. In the lower left side of this window, we find information on the simulation process including whether or not any errors occurred during the process. To the right is an area that has multiple purposes depending on which tab is selected. For this simple simulation, the only information is contained in the Devices tab. The upper part of the window shown as dark gray will be used later to display printed and plotted results.

In the background during the simulation run, PSpice creates an output file (.OUT). It contains bias point information, model parameter values, error messages, and so on. If the simulation fails, you can view the output file to see the error messages. If the simulation completes successfully, PSpice produces a data file (.DAT). This is the file PSpice uses to display the simulation results.

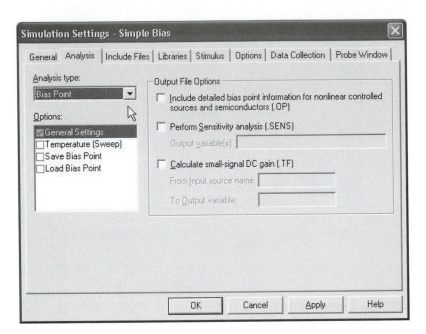

Figure 13 Simulation Setting dialog used to choose the analysis type.

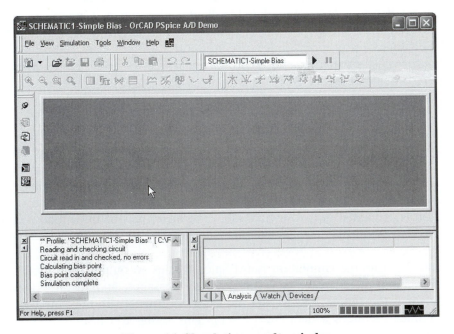

Figure 14 Simulation results window.

2.3   Simulation Results

Click on the button that looks like this [icon], and is the third from the top on the left side list of buttons in the simulation window.   By clicking on this button we open the .OUT file for this simulation, which is created anew each time PSpice is run on this circuit description.   The upper portion of the simulation results window will be filled with a textual description of the results of the simulation run similar to what is shown in Figure 15.   The details of the netlist will be discussed further in SECTION 2.5.   For now we note that R1 is set between the node labeled N00015 and the node N00009, while R2 is set between N00009 and node 0.   The voltage source is connected from node N00015 and node 0.

Scrolling down further in this part of the simulation output window, we can see the results of the nodal analysis as shown in Figure 16.   Node N00009 is listed as 6.6667 volts while node N00015 is listed as 10.0000 volts.   Each of these voltages is referenced to the node labeled 0.   In other words, if we assume that the node labeled 0 is zero volts, then node N00009 is 6.6667 volts above zero while node N00015 is 10 volts above zero.   The fact that N00015 is 10 volts above ground is established by connecting the constant voltage source V1 in the circuit.   PSpice calculates the voltage for N00009 during the simulation.   We can check the N00009 voltage by recognizing that the circuit is a simple resistive divider circuit.   The voltage across node N00009 is calculated as

$$V(N00009) = 9400*10/(4700+9400) = 6.6667.$$

```
**** INCLUDING resistor1-SCHEMATIC1.net ****
* source RESISTOR1
R_R1            N00015 N00009  4.7K
R_R2            0 N00009  9.4K
V_V1            N00015 0 10Vdc


**** RESUMING "resistor1-schematic1-simple bias.sim.cir" ****
.INC "resistor1-SCHEMATIC1.als"



**** INCLUDING resistor1-SCHEMATIC1.als ****
```

resistor1-SC...

Figure 15 Portion of the simulation results showing the circuit netlist.

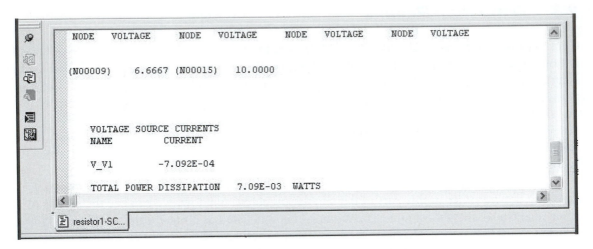

Figure 16 Simulation results with node voltage calculations.

In the output window, PSpice also tells us that the current delivered by the V1 voltage source is –7.092E-04 amps—or –0.7092 milliamps. PSpice calculates this current as

$$I = 10/(4700+9400) = 0.7092 \text{ mA}.$$

PSpice assumes the direction of current flow in the voltage source to be into the positive node of the source. Therefore, since the actual conventional current flow in this circuit is out of the positive node of V1, it is reported as negative.

Finally, PSpice calculates the value of the power supplied by the voltage source V1 and displays it in the bottom of the output window as 7.092 milliwatts. This number is simply the current supplied by the source multiplied by the voltage across the source.

As part of the operation of displaying the output, PSpice sends simulation results information back to the Capture program to be displayed on the schematic diagram. After running PSpice the schematic capture window should look like the one displayed in Figure 17. It may be necessary to click on the **V** button near the top-center of the schematic capture window in order for the voltages to be displayed. This process is generally referred to as back-annotation with the schematic diagram receiving information from the simulation program. Note that clicking on the **I** button causes the back-annotated currents in the circuit to be displayed.

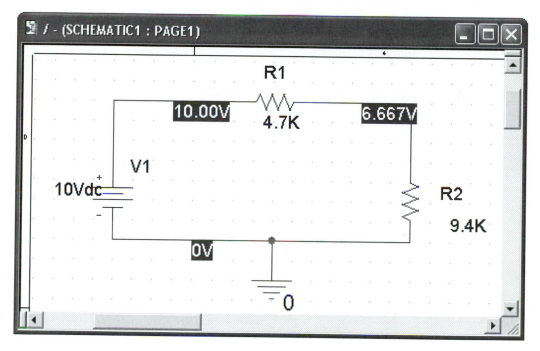

Figure 17 Schematic capture window with PSpice back-annotation.

When viewing results from this and any simulator, we should be sure to do some fundamental calculations of our own to insure that the simulator is providing reasonable results. Various types of errors may cause a simulation to produce results that are out of the domain of the expected solution. A good designer must be able to recognize when a set of results is likely to be inside the domain of correct possibilities.

2.4   *Generating the Simulation File by Hand*

PSpice simulations may also be run directly from a netlist description of the circuit that is generated without the use of the capture software. In some cases of repeated simulations and parameter modifications, it may be easier to run the process using a netlist rather than the schematic capture tool. To understand PSpice simulation file format, we will generate by hand a PSpice description for the circuit of Example 1. Starting with the diagram in Figure 2, we number each of the nodes of the circuit to obtain the circuit shown in Figure 18. We choose to number the node at the bottom of the power source as node zero. (A common practice is to choose the node that has the largest number of connection as the reference node and label it as node 0.) In this case we have used sequential numbering although the newer versions of PSpice allow for arbitrary node numbers or names.

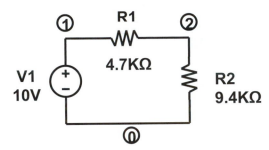

Figure 18 Example 1 circuit with node numbering.

Next we produce a text file using a text editor like MS Notepad that contains a description of the interconnections of the circuit—the netlist—along with PSpice specific commands for the simulation. The netlist begins by describing the circuit components using the format:

<component name> <node +> <node -> <value> <other component parameters>

The component name tells PSpice what type of component we are describing as well as providing a unique identifier for this instance of that type of component. PSpice determines the component type from the first letter of the component name. Table 2 contains a sample list of identifier letters and the corresponding component types available in PSpice.

The voltage source and the two resistors in this circuit are described by the statements:

```
V1  1  0  10
R1  1  2  4.7K
R2  2  0  9.4K
```

Multiplying factors from Table 1 may be used in the value field of the statement in order to save typing while providing the necessary scale factor.

Table 2 PSpice Component Naming Conventions

Component Identifier Letter	Component Type
C	Capacitor
E	Voltage-Controlled Voltage Source
F	Current-Controlled Current Source
G	Voltage-Controlled Current Source
H	Current-Controlled Voltage Source
I	Independent Current Source
K	Inductor Coupling (Transformer)
L	Inductor
M	MOSFET Transistor
R	Resistor
T	Transmission Line
V	Independent Source

Two more statements must be added to the circuit description file in order to perform the simulation. PSpice always assumes that the first line encountered in the circuit description file is a title for the circuit. Also, PSpice needs to have a .END statement as the final line in the file in order to know when the input data is complete. Figure 19 contains a circuit description file for the simple restive circuit of Example 1. Note that the file type used is .cir since it best matches the file type produced in Capture.

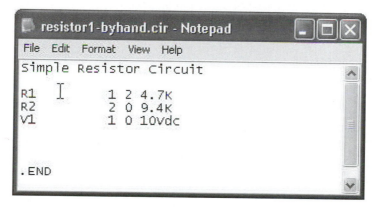

Figure 19 Text file for simulating the circuit in Figure 18.

To simulate a circuit using a file like the one of Figure 19, open the PSpice simulation window (OrCAD PSpice A/D Demo) and click on File>Open. Then in the dialog box, use the pull down menu under Files of Type to specify files of type: Circuit Files (*.cir). Next navigate to, and select, the file we generated containing a circuit description similar to the one in Figure 19.

Finally, select Simulation>Run or the ▶ button to perform the simulation. Results may be viewed by clicking on the View Simulation Output File button 🖅 on the left side of the window. These results should be identical, in calculated values, to those produced when using the Capture tool. The potential advantage in the by-hand technique lies in the ability of the user to rather quickly change the value of circuit parameter values by merely opening the .cir file with a text editor and editing the parameter(s) of interest.

There are a few rules that must be kept in mind when describing circuits using this by-hand method. First, each node must have a dc path to ground. This is necessary because of the way in which PSpice writes and solves the equations that describe the behavior of the circuit. The second connection rule is that each node be connected to at least two circuit components. This precludes leaving a circuit element dangling, unconnected on one end. Although we may leave dangling connection open on a lab test bench, PSpice doesn't allow it. In many cases, a very large resistor on the order of hundreds of megaohms, connected from a node to ground, may be used to produce the effect of leaving a node disconnected. This resistor will have little effect on most circuits that have typical resistance values several orders of magnitude smaller.

At this point we should have a good understanding of how to perform a basic simulation on a multi-node circuit. The rest of this book will describe how to extend these techniques to more complicated circuits and more sophisticated types of simulation.

Chapter 3 Simple DC Circuits

In this chapter we will extend our analysis to go beyond circuits that contain just resistors and independent voltage sources; we will now include independent current sources as well as dependent sources. The analysis type will continue to be a bias point analysis seeking to establish the DC operating point for the circuit.

3.1 Independent Sources

Two types of independent sources are available for use in PSpice simulation. We have already used an independent voltage source in our earlier simulations to model the behavior of a constant voltage source such as an ideal battery. Independent voltage sources produce a constant voltage across their terminals independent of the current flowing through them or the behavior of any other circuit elements or circuit conditions. Another type of independent source is an independent current source. Independent current sources produce a constant amount of current independent of the other elements in the circuit. Schematic models for these sources are available in the parts library named Source with the independent voltage source being named VDC and the independent current source named IDC.

3.2 Dependent Sources

Dependent sources are circuit elements whose behavior is controlled by a current or voltage that appears elsewhere in the circuit. Both dependent voltage source models as well as dependent current source models are available in PSpice. Either of these two sources may be regulated by a controlling voltage or current. Thus, the four available dependent sources are: a voltage-controlled voltage source, a current-controlled voltage source, a voltage-controlled current source, and a current-controlled current source.

Dependent sources are used to model the behavior of physical devices. For example, a voltage-controlled current source is used to model the behavior of a MOSFET that is acting as an amplifier. The general way in which this is done is to use the voltage input to the transistor to determine the amount of current that flows through the transistor. The details of this type of modeling are left as the subject for an electronics course. We will analyze a relatively simple resistive circuit to illustrate the type of configuration in which you may find a controlled source.

Example 2: Analyze the circuit of Figure 20 and determine the operating point voltages and currents through all elements of the circuit. The current in the voltage-controlled current source is set to be i = 0.7 × v_{R5}, where v_{R5} is the drop across R5 as noted in the diagram.

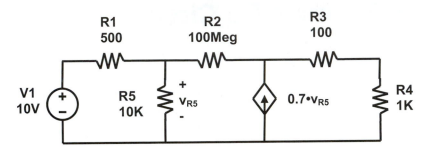

Figure 20 Example circuit with dependent source.

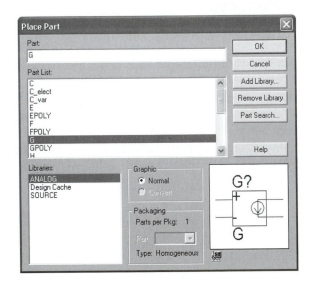

Figure 21 Dependent current source part window.

Solution 2: The PSpice analysis of the circuit is performed in nearly the same way as the bias point analysis of Example 1. First, the circuit is drawn with the schematic capture tool after which it is analyzed by PSpice and the results are reported in a table of node voltage values. The voltage-controlled current source is taken from the analog library and is the part called G as shown in Figure 21. Notice that in its generic form, the part has its current source oriented in the opposite direction from the current source specified in Figure 20. Therefore, after the G part is placed in the circuit, the value of the current source must be adjusted to correspond to the polarity indicated in the circuit drawing.

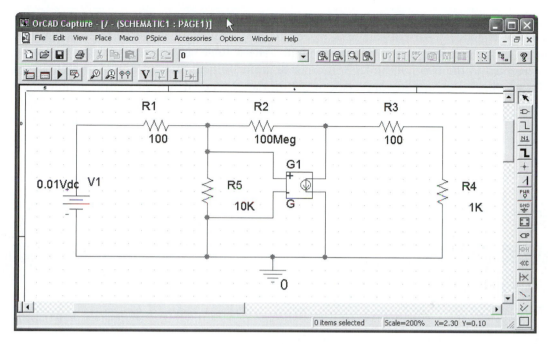

Figure 22 Schematic for circuit of Example 2 containing dependent current source.

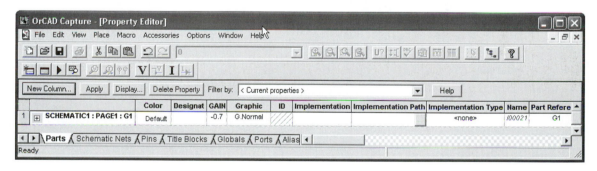

Figure 23 Property Editor for a voltage dependent current source.

Figure 22 contains the schematic capture window for the circuit being analyzed for Example 2. The voltage-dependent current source is placed in the circuit so that the current source can be connected to the nodes of resistor R3 and ground. The voltage sensing leads labeled + and – are then connected to the two nodes of R5 in order to sense voltage drop across R5, which is called v_{R5}. To set the current multiplying factor to 0.7 for the dependent current source, double click in the center of the G1 symbol opening the Property Editor window for the G1 part as shown in Figure 23. Place the cursor in the GAIN field and click. Edit the value of the current source gain to be -0.7. By making the gain multiplier negative, we have effectively reversed the direction of the current source to match the direction of the dependent current source specified in the original problem. Be sure to click the Apply button in the upper left of the window before closing the Property Editor window with the ☒ close button in the upper right corner of the window. (Be sure to close the inner Property Editor window and not the overall OrCAD Capture window.)

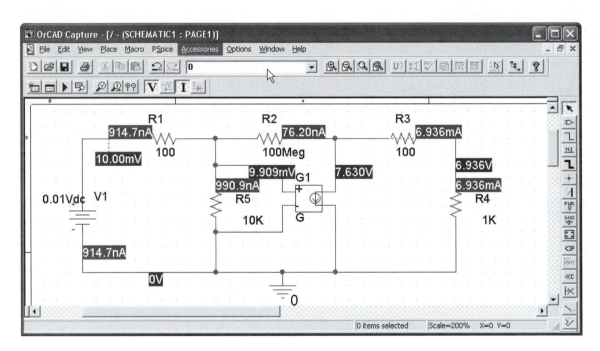

Figure 24 Solution to circuit with voltage-dependent current source.

To set up the simulation, click on the PSpice>New Simulation menu item and choose the Analysis tab. Specify a name for this simulation run under the General tab by filling in the text area Simulation Profile with a unique identifier name. Next, choose the Bias Point analysis from the Analysis Type pull down menu and click OK to complete setup of the simulation profile. To run the simulation, click on PSpice>Run from the menu or click on the Run PSpice button ▶.

Results of the simulation can be viewed by clicking on the $\boxed{\mathbf{V}}$ button and the $\boxed{\mathbf{I}}$ button in the OrCAD Capture window.   The back-annotated values of the node voltages and the branch currents for the circuits will be displayed similar to those in Figure 24.

It is always a good idea to check the results of the PSpice simulation; if not in full detail, we should at least check to see if the results are reasonable. To begin a check, determine if the currents at the nodes abide by Kirchoff's Current Law. First we need to recognize that the current out of the positive side of the voltage source is 914.7 nA. At the node between R1 and R2, the currents are 914.7 nA + 76.20 nA = 990.9 nA, which is correct by KCL. The current through R2 is known to flow from right to left since the voltage on the right node of R2 is higher than the voltage on the left of R2.   Control voltage v_{R5} = 990.9 nA × 10kΩ = 9.909 mV which is the current entering the top of R5 multiplied by the resistance of R5. The current in the voltage-controlled current source G1 is i_{G1} = -0.7 × 9.909 mV = -6.936 mA which is the multiplying factor -0.7 multiplied by the controlling voltage v_{R5}. Two things to notice here are that the minus sign in this equation accounts for the fact that the current we calculate is opposite to the direction of the current source drawn in G1 in Figure 24.   The second important detail is that the multiplication factor 0.7 must have the units of A/V in order to account for the fact that a voltage controls the amount of current output from the source. Checking the current at the node between R2 and R3 we have 6.936 mA – 76.20 nA = 6.936 mA. The seeming discrepancy here is due to the limited precision of the branch currents printed in the schematic. In fact, if we examine the branch currents with several more digits of precision, you find that KCL is also observed at this node. Further evidence of the accuracy of the solution is found when the currents entering node 0 are summed. We leave it to the reader to perform the KCL calculations at this node.

3.3   Thevenin Equivalent Circuits

Thevenin's Theorem states that a linear two-terminal network can be replaced by an equivalent circuit consisting of a voltage source of value V_{Th} in series with a resistance of value R_{Th}, where the value of V_{Th} is the open-circuit voltage of the original network and the value of the resistance R_{Th} is the equivalent resistance at the terminals when the independent sources are turned off. Another way of calculating R_{Th} is to divide the open-circuit voltage by the short-circuit current with all sources acting. A general form of the Thevenin equivalent circuit is shown schematically in Figure 25.

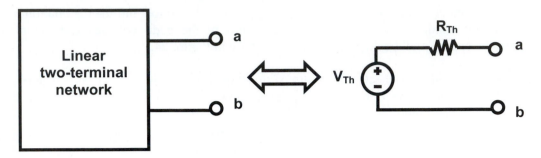

Figure 25 General form of the Thevenin Equivalent circuit.

PSpice can be used to calculate the Thevenin equivalent voltage source and the Thevenin equivalent resistance by first entering the circuit into the schematic capture program and then placing circuit elements at the output that allow us to measure the open circuit voltage and the short circuit current. The Thevenin voltage $V_{Th} = V_{oc}$ and the Thevenin resistance $R_{Th} = V_{oc}/I_{sc}$, where V_{oc} is the open-circuit voltage and I_{sc} is the short-circuit current at terminal pair **a-b**.

Example 3: Determine the Thevenin equivalent circuit for the network shown in Figure 26.

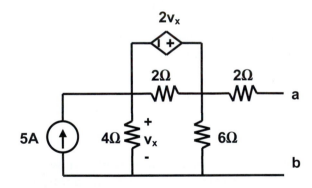

Figure 26 Determine Thevenin equivalent circuit at nodes a-b.

Solution 3: Draw the schematic in the OrCAD Capture program as shown in Figure 27. The element E1 is used to model the voltage-controlled voltage source with the internal gain factor being set to two in the same way the gain was changed in Solution 2 above. When we attempt to simulate the circuit behavior with PSpice, we receive an error indicating that the resistor R3 has only one other branch connected to it. The large circle at node **a** is the back-annotation from PSpice that indicates that this open-ended element is illegal for PSpice. To overcome this problem we connect a very large resistance across the nodes labeled **a-b**. We use a resistance of 1000 Gigaohms as shown in Figure 28, which models an open circuit and satisfies the PSpice branch connection requirement while allowing us to measure the open circuit voltage across **a-b**. From Figure 28 you can see that the voltage V_{oc} across R5 is 20 V.

Short circuit current I_{sc} is measured by placing a wire between node **a** and node **b** and examining the current flowing in the wire. Running the PSpice simulation produces the results in Figure 29, specifically showing $I_{sc} = 3.333$ A. Dividing the open circuit voltage by the short circuit current produces the Thevenin equivalent resistance $R_{Th} = 6$ Ω.

Thevenin equivalent circuits are very handy when performing several different types of circuit analyses. For example, when you need to know the effects of a variation in one specific circuit element, you should develop the Thevenin equivalent for the rest of the circuit in the absence of the varied element. Then we can determine the voltage and current in the target element by analyzing the circuit consisting of the Thevenin equivalent circuit in series with the target element. Repeated calculations for this three element circuit are likely to be much less tedious than for the entire circuit.

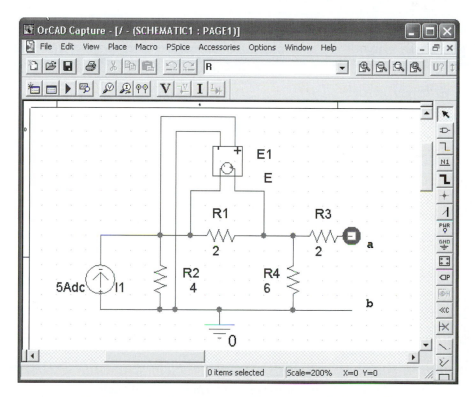

Figure 27 PSpice circuit to calculate Thevenin equivalent of Figure 26.

3.4   Norton Equivalent Circuits

Another form of an equivalent circuit which is useful in certain analysis situations is called a Norton equivalent circuit. Norton's Theorem states that a linear two-terminal network can be replaced by an equivalent network consisting of a current source I_N in parallel with a resistance R_N, where the value of I_N is the short circuit current through the terminals and R_N is the equivalent resistance looking back into the network when the independent sources are turned off. The general form of the Norton Equivalent circuit is shown in Figure 30. The value of R_N is the same as the value of R_{Th}.

To find the Norton equivalent circuit for the network of Figure 26, we perform the same two analyses as when finding the Thevenin equivalent. That is, use the results shown in Figure 29 to determine the value of the I_N as 3.333 A. while from Figure 28 the value of $R_N = V_{oc} / I_{sc} = 6\ \Omega$ Thus, in Figure 31 you see two equivalent circuits, the Thevenin equivalent and the Norton equivalent, for the circuit shown in Figure 26.

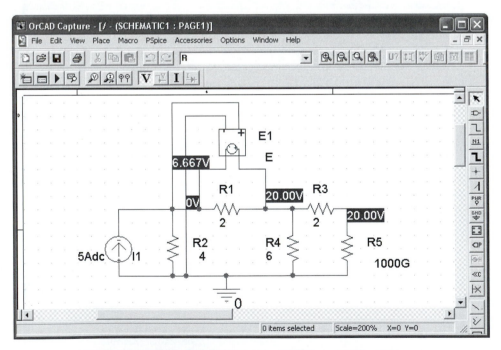

Figure 28 Large resistance at terminals to calculate V_{oc}.

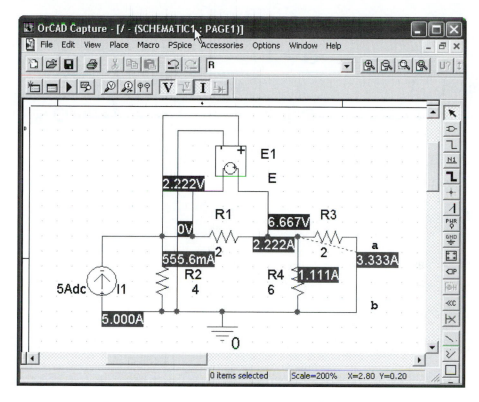

Figure 29 Wire across terminals to calculate I$_{sc}$.

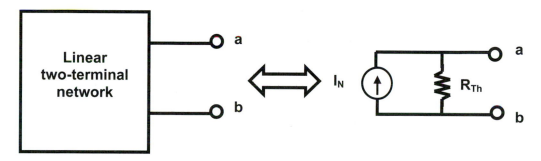

Figure 30 General form of the Norton Equivalent circuit.

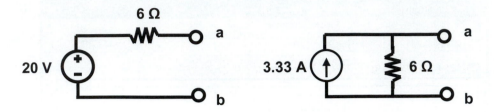

Figure 31 Thevenin and Norton equivalent for the circuit of Figure 26.

Chapter 4 Other DC Analyses

PSpice is capable of performing several additional types of DC analysis beyond the simple operating point analysis discussed in Chapters 2 & 3. In this chapter we will explore how to use PSpice to perform simulations that allow circuit designers to develop circuits tolerant of variations in expected operating conditions and component values.

4.1   DC Sweep Analysis

In many situations during the circuit design process it is necessary to analyze a circuit over a range of DC voltages and currents. Some circumstances require that one or more output voltages or currents be described as a function of inputs. To perform these analyses in an efficient manner, PSpice has a built in capability to sweep specified input sources while tracking output values. The results can be plotted or printed so that the designer has a thorough understanding of the full range of DC behavior of the circuit. Remember, the DC behavior of the circuit is defined by the situation where the capacitors are open circuits, the inductors are shorts, and all DC sources are operative.

Example 4: For the circuit of Figure 32, determine the voltage drop across the nodes labeled v_{out} along with the current through R5, when the voltage produced by the source labeled v_{in} varies from zero to 12 volts.

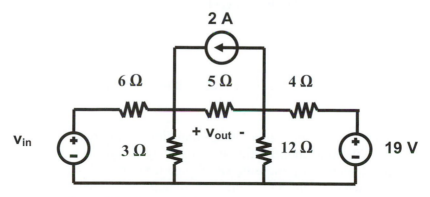

Figure 32 Calculate DC voltage v_{out} as a function of v_{in}.

Solution 4:   To solve the problem, enter the schematic in Figure 32 into OrCAD Capture as shown in Figure 33. Notice that the designator **+ vout -** in the schematic in Figure 33 is simply a label that does not have a connection to the nodes in the circuit. A label can be entered in a schematic by clicking on the Place Text icon 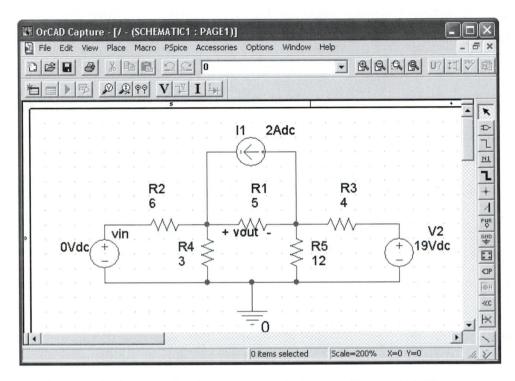 on the locating the cursor on the schematic, and then typing in the desired text.

Figure 33 Analyze the circuit by sweeping the input voltage and plotting vout.

The voltage source vin, which is to be swept during the simulation, is initially set to zero volts dc. The dc value assigned to vin is somewhat arbitrary since PSpice will recognize that this is to be a dc sweep analysis and will use the initial value declared in the Simulation Profile that is set in the next step.

Once the circuit has been drawn in Capture, the Simulation Profile must be set by first going to the menu at PSpice>New Simulation Profile and providing a name for this profile. Next, you must specify the type of analysis by choosing DC Sweep from the pull-down menu in the top left of the dialog box as shown in Figure 34. In the same dialog box, set the Sweep Variable Name to *vin* in order to sweep the value of the voltage source. Finally, set the Sweep Type to *Linear*, the Start Value to *0*, the End Value to *12*, and the Increment to *0.1*. Click on OK and you are nearly ready to perform a simulation.

In order to output the results of the simulation, you must indicate to PSpice which values you would like to display and in what manner you would like to display them. For this example, we will begin by displaying a plot of both vout and I(R5) versus vin. The notation *I(R5)* designates the current flowing in the element R5. To generate these plots, we go back to OrCAD Capture and specify locations to measure this voltage and current. A voltage difference marker is put in place by clicking on the Voltage Difference Marker button and then placing the marker symbols at the nodes of interest as shown in Figure 35. Similarly, a current marker is placed in the circuit by clicking on the Current Marker button and placing the current marker symbol on the branch of interest—in this case R5 as shown in Figure 35. The symbols labeled V+ and V- are generated by the Voltage Difference Marker button and indicate to PSpice that the voltage drop across R5, in the direction indicated by the + and – signs, will be plotted. The marker attached to the left side of R5 indicates that the current through R5, in the direction from left to right, will be plotted.

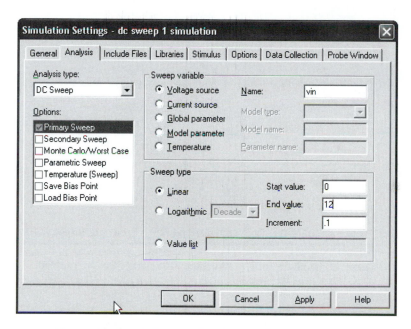

Figure 34 Simulation Setting for a DC sweep analysis.

Clicking on the Run ![button] button (or on PSpice>Run from the menu bar) starts the simulation and produces 121 data points for both the requested voltage and the current, and subsequently plots the data in the results window. Using the voltage and current markers causes the plot of these two data sets to appear with no further user intervention required. A portion of the output window showing the data plots is shown in Figure 36. You should recognize that the output curves are straight lines as would be expected for both voltage and current in this simple linear circuit.

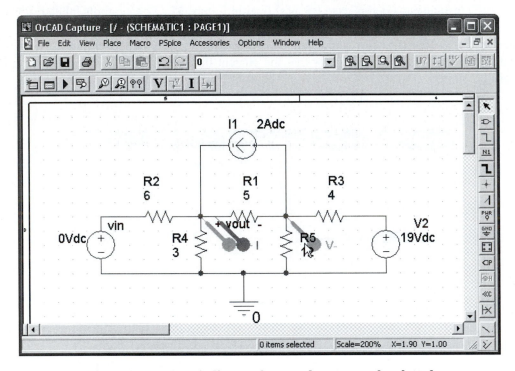

Figure 35 Markers indicate voltage and current to be plotted.

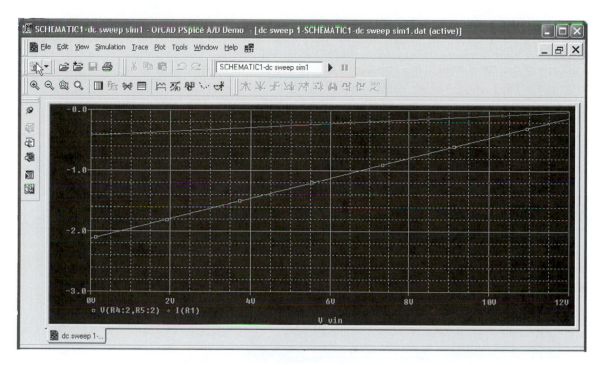

Figure 36 Plot of vout and I(R5) vs. vin for Example 4 circuit.

The output seen in Figure 36 is plotted by software called Probe. This plotting software is very powerful and will be discussed at more depth in Chapter 11. For now we will do a few simple manipulations with Probe. Start by clicking on the Toggle Cursor button   and notice that a dashed line cursor appears in the plot area and that a Probe Cursor data box appears in the bottom right of the PSpice output window. In the lower left portion of the plot is a legend showing the color and symbol used in the curve representing the voltage and the current in your plot. You should also notice that there is a dashed box around one of the two symbols. This indicates that the cursor being shown is tracking that particular output value. Click on the symbol representing the voltage. Voltage vout and the corresponding value of vin at specific points on the plot is shown on line A1 of the data box. We will use the data shown in this data box to get a more precise value of the voltage or current at various places on the plotted curves.

An example of the usefulness of this type of analysis capabilities can be seen in the following hypothetical situation. Suppose that a circuit was produced with the configuration shown in Figure 32 and that a value for vin must be found so that vout was equal to zero. From the original plot shown in Figure 36 you can see that at the maximum input plotted the output does not reach zero. Therefore, we need to re-run the simulation after extending the range of the sweep. Select PSpice>Edit Simulation Profile from the menu and change the End Value in the Sweep Type settings to 19 instead of 12 (see Figure 34). Click the Run button and the plot shown in Figure 37 will appear in the output window. From the figure you can see that the voltage vout reaches zero when vin is about 12.7 V. To be more precise about the value of vout, you can use the left and right arrow keys to move the cursor so that the vout value is as close to zero as possible. In this example, you will find that the closest to zero that you can get for vout is either -2.193 mV or 2.193 mV. The corresponding values for vin are 12.737 V and 12.763 V, respectively. Since this is a linear system, you can extrapolate to obtain the value of vin = 12.750 V when vout = 0 V.

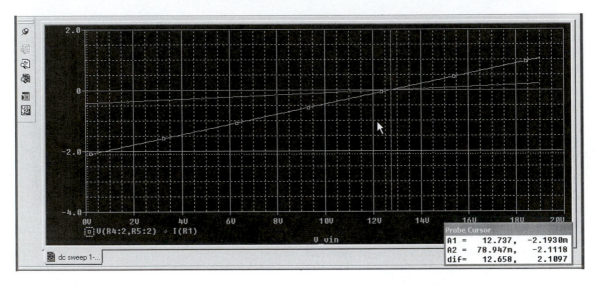

Figure 37 Output plot over extended range with cursor displayed.

The dc sweep analysis is used in a number of different situations during a circuit design process. Designers use graphical techniques to obtain a visual understanding of the analysis and of the output changes that can be expected when circuit parameters change. The next section will further explore the ability of PSpice to support design decisions and tradeoffs.

4.2   DC Sensitivity Analysis

PSpice provides the Sensitivity or .SENS Analysis, to help the designer predict a circuit's response to changes in component values. Typically this component parameter change is caused by variations in the behavior of the circuit element, although the Sensitivity Analysis may also be used to help you zero-in on an appropriate parameter value during the course of the initial circuit design. In a Sensitivity Analysis, PSpice calculates the dc small-signal sensitivity of specific output variables with respect to each circuit parameter. Since the Sensitivity Analysis is a dc analysis, only resistors, voltage sources, current sources or models containing these elements are considered. To perform a Sensitivity Analysis you must select the voltages or currents whose sensitivities are to be calculated. Only currents flowing in voltage sources may be specified for a sensitivity analysis. Since this analysis calculates the sensitivity of the selected variable(s) with respect to each of the dc circuit elements, there is the possibility of generating huge amounts of data for a relatively small circuit. The user must make careful choices in order to avoid long simulation times and large data files. The following example describes the basics of producing and applying a sensitivity analysis.

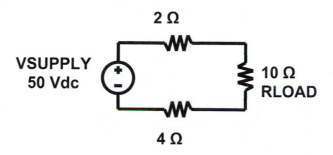

Figure 38 Circuit for Example 5.

Example 5: The circuit in Figure 38 is a crude model for a dc power transmission line. Determine the sensitivity of the voltage and current delivered to the load 10 Ω load resistance when variations occur in the circuit's resistances.

Solution 5: The circuit shown in Figure 39 is drawn in OrCAD Capture as a model of the circuit of Figure 38. The voltage source VDUMMY is used to allow us to monitor the current flowing through the load resistor RLOAD during the sensitivity analysis. (Remember, only currents flowing in voltage sources may be monitored and compared during a sensitivity analysis). In order to have no effect on the behavior of the original circuit, VDUMMY is set to zero volts, making the source effectively a short circuit from the resistor to the ground node. Also included in this drawing is what Capture defines as an *off-page connector*. This connector symbol allows us to label a node and subsequently refer to that node by name—in this case *Sig1*.

To set up the Sensitivity Analysis, click on PSpice>Edit Simulation Profile and use the pull down menu to set the Analysis Type to be Bias Point as in Figure 40. Also, in the Output File Options portion of this dialog window, check the box labeled Perform Sensitivity Analysis (.SENS). Enter the variable names V(Sig1) and I(VDUMMY) in the text box labeled Output Variable(s) indicating which circuit variables should be monitored and reported during the simulation. The variable name V(Sig1) specifies the voltage drop from the point labeled Sig1 to the reference

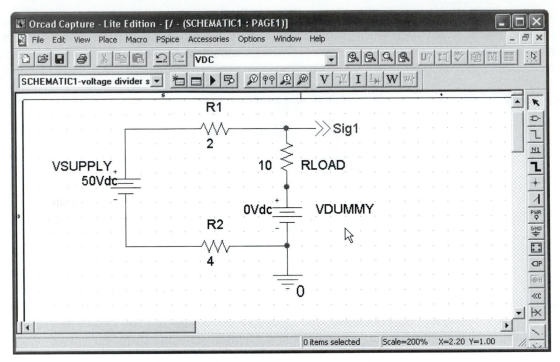

Figure 39 Schematic for circuit in Figure 38.

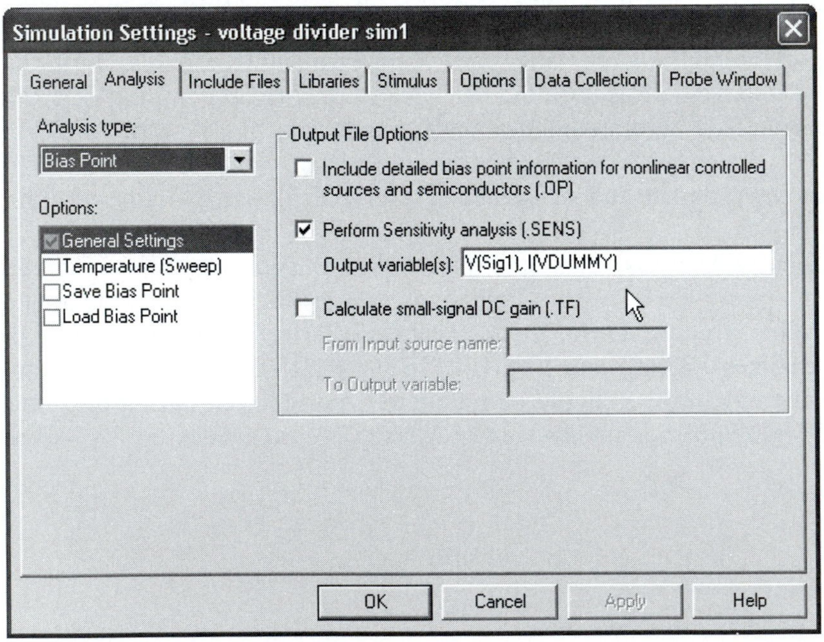

Figure 40 Simulation Profile setting for performing a Sensitivity Analysis.

node (ground). I(VDUMMY) is a variable name representing the current flowing through the voltage source VDUMMY in the direction from the positive node to the negative node. Once the setup is complete, click on the Run button to have PSpice perform the simulation.

Results of the simulation can be viewed by clicking on the View Simulation Output button in the PSpice window. Scrolling down in the output window takes you to the data shown in Figure 41, the first section of which contains data for the variable V(Sig1). The first element listed is R_R1. (Note: R_R1 is the internal name that PSpice uses to designate resistor R1. The R_ designates the element as a resistor as required by PSpice. Similarly, V_ is used as a preface for voltage sources, L_ for inductors, etc.) Next to the element name R_R1, the element value 2.000E+00 is displayed in scientific notation format representing the number 2.000×10^0 or 2.000. Column three contains the element sensitivity of -1.953 volts/unit for R1. This means that the voltage at node Sig1 will decrease by 1.953 volts for each ohm of increased resistance in R1. If R1 is increased by a factor of three to become 6Ω, the voltage at node Sig1 will decrease by a factor of 5.859 volts to 25.391. The Normalized Sensitivity values are calculated by multiplying each Element Value by the Element Sensitivity value and dividing by 100. For every 1% change in the element value, there will be a change in the value of Sig1 in the amount of the Normalized Sensitivity. For example, if the voltage source VSUPPLY is increased by 1%, the output voltage Sig1 will increase by 0.3125 volts.

Similar results are shown for the behavior of the current I(VDUMMY) flowing in the resistance RLOAD. The amount of the change of current as a function of the change in element value is shown in Figure 41.

DC Sensitivity is calculated with respect to the DC operating point of the circuit. PSpice calculates the difference in an output variable by perturbing each element parameter independently. Since the circuit is a linear circuit, the principle of superposition applies and all of the sensitivity results can be overlaid. For example, if R1 is increased by 5% and R2 is decreased by 3%, the overall effect of these changes is found by summing each of the individual effects. In this case, the voltage at node Sig1 will increase by 0.0391 volts.

```
DC SENSITIVITIES OF OUTPUT V(SIG1)

             ELEMENT         ELEMENT         ELEMENT         NORMALIZED
             NAME            VALUE           SENSITIVITY     SENSITIVITY
                                             (VOLTS/UNIT)    (VOLTS/PERCENT)


             R_R1            2.000E+00       -1.953E+00      -3.906E-02
             R_RLOAD         1.000E+01        1.172E+00       1.172E-01
             R_R2            4.000E+00       -1.953E+00      -7.813E-02
             V_VSUPPLY       5.000E+01        6.250E-01       3.125E-01
             V_VDUMMY        0.000E+00        3.750E-01       0.000E+00


DC SENSITIVITIES OF OUTPUT I(V_VDUMMY)

             ELEMENT         ELEMENT         ELEMENT         NORMALIZED
             NAME            VALUE           SENSITIVITY     SENSITIVITY
                                             (AMPS/UNIT)     (AMPS/PERCENT)


             R_R1            2.000E+00       -1.953E-01      -3.906E-03
             R_RLOAD         1.000E+01       -1.953E-01      -1.953E-02
             R_R2            4.000E+00       -1.953E-01      -7.813E-03
             V_VSUPPLY       5.000E+01        6.250E-02       3.125E-02
             V_VDUMMY        0.000E+00       -6.250E-02       0.000E+00
```

Figure 41 Results of the Sensitivity Analysis for two selected output variables.

Users must be careful when interpreting Sensitivity Analysis results. For example, suppose that we increased the size of R1 to be 100 Ω. The results in Figure 41 would indicate that the voltage at Sig1 should decrease by 97.65 volts. This of course is impossible since the supply voltage is only 50 volts. Thus it is important to note that the Sensitivity Analysis results are only valid over a narrow range of values near the operating point of the original circuit. Users must be careful to check the reasonableness of the results before they apply them.

Sensitivity Analysis is very useful when trying to find a worst-case scenario for circuit operation. By finding the most sensitive elements and adjusting their values to decrease sensitivity, the designer can produce a more robust design. Alternatively, if adjustment is not possible due to specification constraints, the designer may stipulate that high-quality components must be used in these locations in order to ensure circuit reliability.

4.3   *Simulating Resistor Tolerances*

Calculations and simulations used to predict circuit behavior are based upon mathematical models of circuit elements and configurations. In most cases, ideal components are used to represent the behavior of real-world components. Unfortunately, real-world components do not always behave in the same manner as the ideal component. In order to better model real-world operation, designers typically analyze their circuits over a range of possible component values. Ranges are chosen based on expected variations in component values which can generally be predicted by component manufacturers. For example, resistance values are specified by the manufacturer to be within a certain tolerance of the nominal value; i.e., a resistor may be sold as a 10 Ω resistance plus or minus 5%. The value +/-5% is said to be the tolerance of the part. Typically, parts that have tighter tolerances are more expensive. In the next example, we will show how to simulate systems containing components with known tolerance values, and how we can produce a worst-case analysis for the system.

Example 6:   Change the model for the RLOAD resistor in Figure 38 so that it more accurately reflects the behavior of a resistor that has a 10% tolerance. Simulate the circuit behavior to get a better understanding of the range of possible circuit responses that can be expected.

Solution 6:   We will modify the model for the RLOAD resistor of Figure 39 by changing element parameters with the property editor. Double-click on the resistor RLOAD to open the property editor. Move the slider at the bottom right of the window so that the column labeled TOLERANCE is viewable. Enter the value 10% as shown in Figure 42 and then click on the Display button in order for the tolerance property to be displayed in the schematic. Click Apply then close the property editor by clicking on the X in the upper right corner of the inner window after which the schematic shown in Figure 43 should appear.

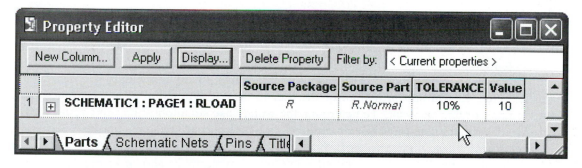

Figure 42 Property Editor used to modify resistor tolerance parameter.

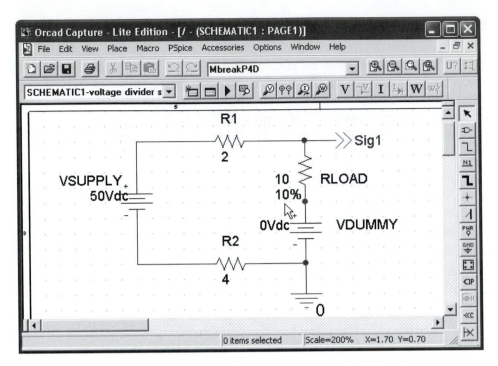

Figure 43 Schematic showing RLOAD resistor with 10% tolerance.

To perform the circuit analysis based on the 10% tolerance in the resistor property, we must set up the analysis by modifying the simulation settings. Pull down the menu item PSpice>Edit Simulation Profile and set the analysis type to be DC Sweep. Set the sweep variable to be VSUPPLY with the Start value and End value set to 50, and the Increment set to 1. Next, click on the check box labeled Monte Carlo/Worst Case, which will change the Simulation Settings window to look like the one in Figure 44. Click on the radio button Worst-case/Sensitivity and set the output variable to be V(Sig1). Pull down the menu item next to the phrase "Vary devices that have" and select only DEV as shown in the figure. This indicates that only devices with tolerances parameters specified in their property table will be varied during this analysis. Next, click on the More Settings button in the Simulation Settings window of Figure 44. The window shown in Figure 45 will appear and allow the user to choose a function to be performed on the output variable for this analysis. To determine the worst-case situation(s) for our circuit, we must analyze the circuit and determine the maximum worst-case output voltage at node Sig1 as well as the minimum worst-case output voltage. First choose "the maximum value (MAX)" in the Find: pull-down menu in the Worst-Case Output File Options window as shown in Figure 45. Click

OK twice and run the PSpice simulation. Click on the View Output button 🔲 to see the simulation results.

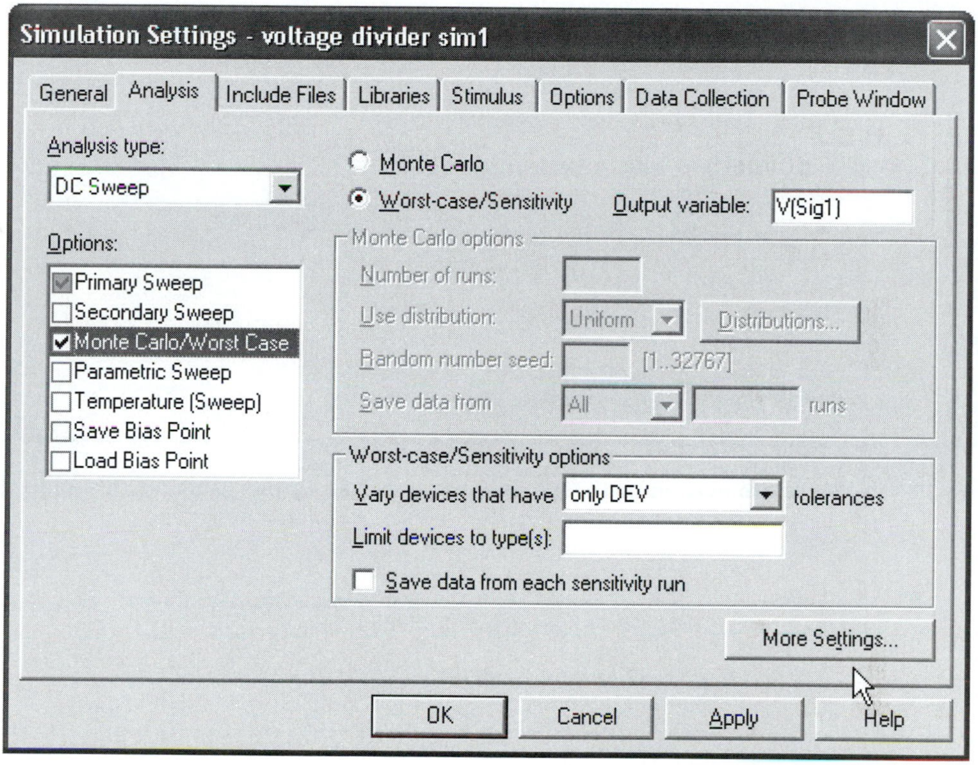

Figure 44 Simulation Setting window used to analyze resistor tolerance variations.

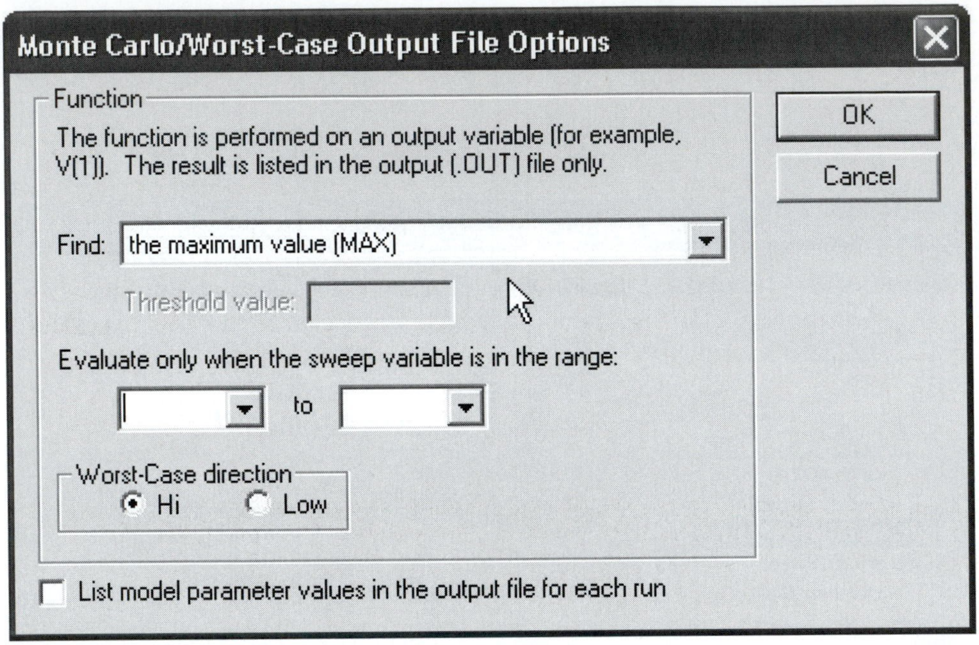

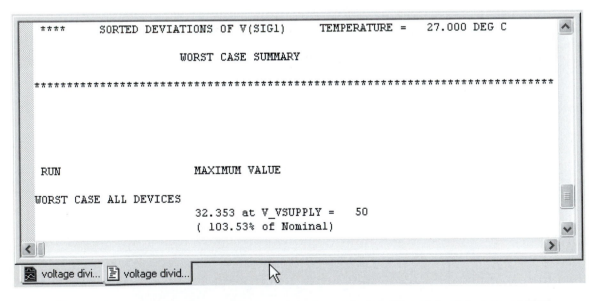

Figure 45 Select MAX to determine the maximum worst-case output value for this analysis.

Figure 46 Output from the maximum worst-case analysis with resistor tolerance at 10%.

Relevant portions of the output file with values for Sig1 are shown in Figure 46. The values output indicate that the voltage at node Sig1 has increased to 32.353 when the value of RLOAD is increased by 10% to be 11Ω. You can easily check this by hand calculating the voltage across RLOAD as SIG1 = 50*11/(2+4+11) = 32.3529 V.

To complete this simple worst-case analysis, we must analyze the situation when RLOAD is reduced by 10%. Repeat the steps that get you to the window of Figure 45 and this time choose "the minimum value (MIN)" in the Find pull-down menu and Low for the Worst-Case direction as shown in Figure 47. Running the simulation produces the results shown in Figure 48. In this simulation, the output voltage at Sig1 is 30 V, which again can be verified by calculating the voltage drop across RLOAD using the resistor voltage divider equations.

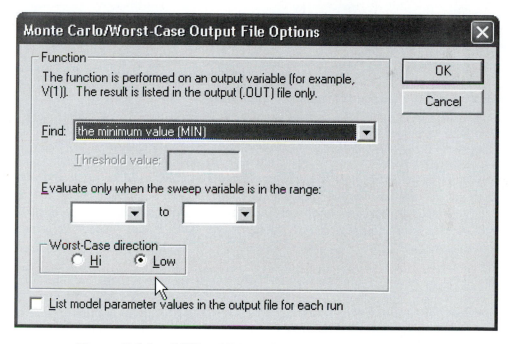

Figure 47 Select MIN and Worst-Case direction Low to find the minimum worst-case situation.

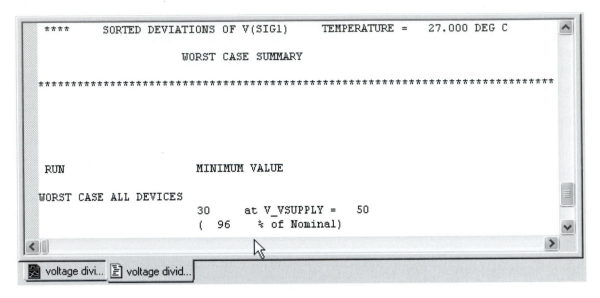

```
****        SORTED DEVIATIONS OF V(SIG1)      TEMPERATURE =    27.000 DEG C

                    WORST CASE SUMMARY

************************************************************************************



  RUN                   MINIMUM VALUE

WORST CASE ALL DEVICES
                    30       at V_VSUPPLY =    50
                    (  96     % of Nominal)
```

Figure 48 Output for minimum worst-case simulation.

The result of the overall analysis shows that if the RLOAD resistor varies between +/-10% of its nominal value of 10Ω, the output voltage at node Sig1 will vary between 32.353 V. and 30 V. When considering the result of this analysis, the designer now has to decide whether the variance in output voltage is acceptable, or if a design change is required. One possible design change might be to select a lower tolerance resistor for RLOAD. This generally is a more expensive alternative and needs to be considered very carefully for devices that will be mass produced. Another alternative is to attempt to modify the circuit so that it is less susceptible to variances in component tolerance.

The power of a worst-case analysis can be better seen when we consider multiple element tolerance variations in parameter values. For example, if you attach a 10% tolerance to each of the resistors of the circuit in Figure 43 and run the two analyses for minimum and maximum worst-case output voltages, the results shown in Figure 49 and Figure 50 will be obtained. These windows show that if all resistors vary by +/-10% then the worst-case maximum output at node Vsig1 will be 33.537 V. This situation occurs when the RLOAD is increased by 10% and the other two resistors are decreased by 10%. Similarly, the minimum worst-case output voltage at node Vsig1 will be 28.846 V. when RLOAD is reduced by 10% and the other two resistors are increased by 10%. You can easily see that worst-case analysis for circuits containing multiple elements can be fairly tedious when done by hand. PSpice is a valuable tool to reduce analysis time for this situation, allowing the designer to spend more time considering design alternatives.

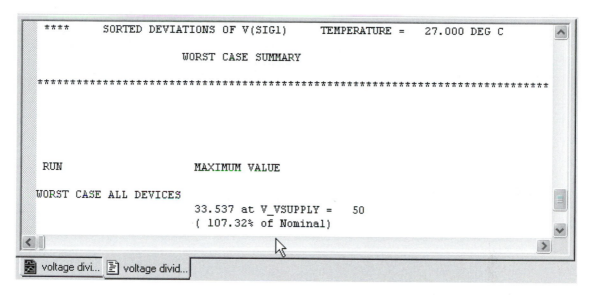

Figure 49 Maximum worst-case analysis with all resistors having a 10% tolerance.

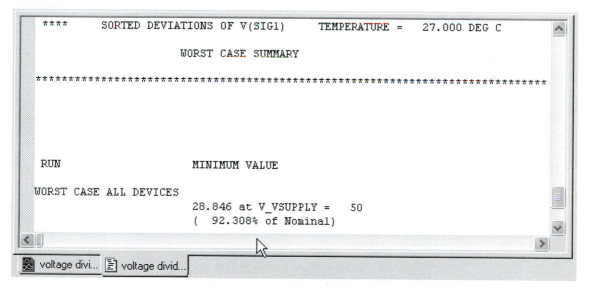

Figure 50 Minimum worst-case analysis with all resistors having a 10% tolerance.

Chapter 5 Operational Amplifiers

Operational amplifiers (op amps) are complex electronic circuits that are important components of many analog circuit designs. Many designers use op amps in much the same way they use more basic components like resistors, voltage sources, capacitors, etc. In courses on circuit design, we are not concerned with the internal circuitry of the op amp or the integrated circuit containing the op amp. Instead, we will use the well-defined and well-characterized devices as building blocks. We treat them as multi-port networks that have certain characteristics at their input and output ports. Other circuit elements are attached to the model for the op amp and the circuit analysis proceeds in much the same way as that of a typical network containing components we have already discussed.

Op amps are generally modeled in one of three different ways in PSpice. The simplest model is one that uses resistors and dependent sources to model the basic behavior of the op amp. PSpice also has a model for several commercially available op amps built into its basic library of parts. The third mechanism for modeling op amps in PSpice is to build a detailed circuit model using individual components to describe the overall amplifier circuit behavior. Models of this type are generally enclosed in a PSpice construct called a *subcircuit*. Built-in models are basically PSpice subcircuits that exhibit the behavior of commercially available components.

5.1   Simple Op Amp Model

An operational amplifier can be modeled with the equivalent circuit shown in Figure 51. The input section of this equivalent circuit consists of the resistor R_i which models the resistance across the two input terminals labeled + and − in the actual op amp. Resistor R_0 models the output resistance seen at the output terminal of the op amp. The voltage amplifying behavior of the op amp is emulated by the voltage-dependent voltage source which produces an output of $A \cdot v_d$. Amplification factor A adjusts the output voltage in direct proportion to the input voltage v_d. This model can be used to predict the simple behavior of circuits containing op amps.

Example 7:   For an amplifier circuit containing an op amp as shown in Figure 52, determine the output voltage across the load resistor when the input signal is set to 0.5 V. The op amp has an input resistance $R_i = 100$ kΩ, output resistance $R_0 = 500$ Ω, and an amplification factor of A = 10,000.

Solution 7:   Figure 53 contains a circuit used to model the amplifier in Figure 52.   The voltage-dependent voltage supply E1 is set to have a gain of -10,000 in the property editor.   A negative gain is used to account for the polarity of the input voltage of the op amp being opposite that inherent in the element E1.   We could also have reversed the wiring of E1 to the rest of the circuit in order to reflect the reversed polarity, but that would have made the schematic unduly messy—using the negative gain is preferred.   Resistor R1 models the 100 kΩ input resistance of the ideal op amp while resistor R4 models the 500 Ω output resistance.   Voltage source VIN is set to the input voltage specified in the question 0.5 V.

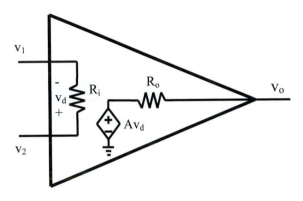

Figure 51 Idealized model for the operational amplifier.

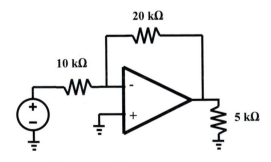

Figure 52 Simple amplifier circuit.

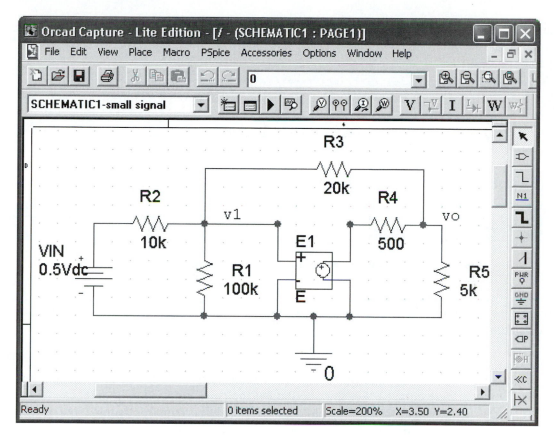

Figure 53 Schematic for the simple amplifier circuit.

Use a bias point analysis to determine the output voltage for the given input voltage. Click on the Run button and examine the results from the PSpice analysis as shown in Figure 54. We can see that the output voltage at node N00195 (as numbered internally by PSpice) is calculated to be -0.9996 V by PSpice. Calculating the output voltage by hand using the simple op amp model and KCL at the two nodes v_1 and v_o produces the following equations:

$$\frac{0.5 - v1}{10 \times 10^3} = \frac{v1}{100 \times 10^3} + \frac{v1 - vo}{20 \times 10^3}$$

and

$$\frac{vo}{5 \times 10^3} = \frac{-10000v1 - vo}{500} + \frac{v1 - vo}{20 \times 10^3}$$

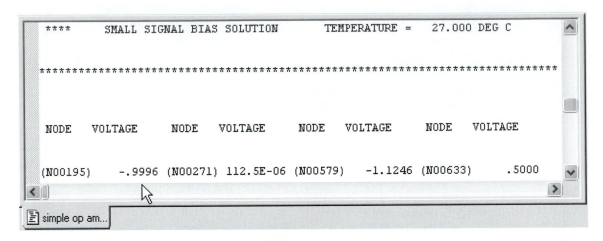

Figure 54 PSpice bias point output for the simple amplifier of Figure 52.

Solving these equations simultaneously produces the value v_o = -0.999 V, which matches the value calculated by PSpice.

PSpice also allows you to exam the input current drawn from the signal source, in this case VIN. The value of the current supplied by sources is automatically included in the output file a few lines below the voltage outputs. For this simulation, the current provided by VIN can be seen to be 0.7498 μA. It is typical for op amps to draw very little current from signal sources making them ideal for sensing signals from sensors and other sources that have very low drive capabilities.

From this result it can be seen that the amplification factor for this amplifier is approximately -2. Checking this by changing the value of VIN to 0.75 V and re-simulating produces the output voltage v_o = -1.4995 V. As expected, this value is -2 times the input voltage. One final check of the amplification factor is performed by setting VIN = -0.75 V. This produces an output voltage of 1.4995 V which shows that the value and the polarity of the output voltage are correctly calculated by PSpice.

The op amp model we have been using thus far is for a non-ideal op amp. In most commercially available op amps, the input resistance is much higher, the gain is much larger, and the output resistance is much lower. If we change the values of these parameters in the model used in the amplifier of Figure 53 so that R_i = R1 = 500 MΩ, $A = 10^8$, and R_0 = R4 = 0 Ω, we will see a slightly different result in the bias point analysis. (Note that PSpice does not allow a resistance value to be set to zero. Instead we set the resistor to be 0.0000001 Ω, which for all intents and purposes is zero ohms yet the value avoids division by zero when used in PSpice's numerical routines.) With VIN = -0.75 V, PSpice calculates the output to be 1.5 V with an input current of 0.750 μA.

5.2   Library Models for Op Amps

The demonstration version of PSpice contains a rich set of device models in the EVAL library. These parts range from models of well-characterized, often-used transistors, to complex digital circuits containing many transistors, resistors and capacitors. Among the set of EVAL parts are several models for commercially available operational amplifiers. These models are more than the simplified idealized model seen in Example 7. Circuit elements in the commercial models allow the designer to simulate the bias point behavior of the op amp as well as the frequency domain and time domain behavior—subjects of latter chapters. One of the most often used op amps in practice is the μA741 op amp which is contained in the EVAL library. We will use it here and in a future example to demonstrate the usefulness of the built-in models.

Example 8: Use the PSpice library model for the widely used commercial op amp named the μA741 op amp to simulate the behavior of a real-world op amp in the amplifier circuit of Figure 52. Determine the voltage gain as well as the input current from the source when V1 is set to 0.5 V.

Solution 8: Open a new project to enter the schematic and simulation parameters for this example circuit. Start to build the schematic by clicking on the place parts button 🔲 in the menu on the right side of the schematic. If the EVAL library is not present in the libraries area of the Place Part window, click on the Add Library… button. Add the EVAL library by clicking on the file eval.olb and then clicking on open (or double-click on the file name eval.olb). Highlight the EVAL library name in the Libraries area and scroll down in the Part List to the very bottom where you will find the μA741 part. Double-click on the part name and add the op amp to the schematic drawing. Add the other parts until your drawing looks similar to that of Figure 55.

In the drawing symbol for the μA741 part, the wires labeled +V and –V must be connected to power supplies of +15 V and -15 V, respectively, in order for the op amp model to function properly. As in Example 5, we have used the Place Off-Page Connector button to include a symbolic connector and reduce the wire clutter in the schematic. Voltage sources V2 and V3 are used to provide the power supply voltages for the μA741 op amp. Connectors +VCC and –VCC are placed at op amp nodes labeled +V and –V, respectively. You will also note that the μA741 op amp model shows numbers for each of the wires attached to the op amp drawing symbol. These numbers correspond to the pins on the standard μA741 op amp chip. It is again important to remember to use the PSpice ground nodes labeled "0" as the common nodes for both the signals as well as the power supplies. Inputs labeled OS1 and OS2 can be used to provide a DC offset voltage, but are not used in this circuit and thus are left open.

Running the PSpice analysis generates output as shown in Figure 56. The output node is N00348, which has the voltage -0.9983 V for this simulation. Other voltages displayed in this figure include voltages that appear at nodes internal to the μA741 op amp model. For example, the node labeled X_U1.6 with voltage 125.8 μV is a node numbered "6" in the subcircuit that describes the μA741 op amp model. We will defer a discussion of subcircuits and their description to the next section.

The input current drawn from the signal source V1 is 0.5 μA, which can be seen from Figure 57. In this figure we can also see values for the current drawn from the power supplies with V2 and V3 each providing 1.667 mA.

You should note that the overall voltage gain for this amplifier is approximately $A_v = 2$. This follows from the fact that the gain can be calculated to be $A_v = R_{FEEDBACK}/R_1$.

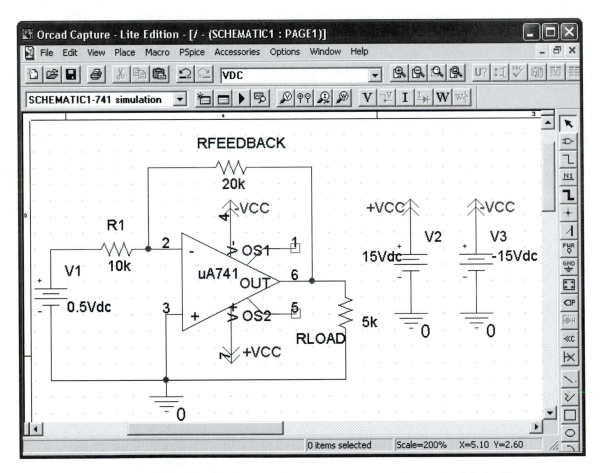

Figure 55 Amplifier schematic using the 741 op amp model from the PSpice library.

```
****      SMALL SIGNAL BIAS SOLUTION       TEMPERATURE =   27.000 DEG C

******************************************************************************

 NODE   VOLTAGE     NODE   VOLTAGE     NODE   VOLTAGE     NODE   VOLTAGE

( +VCC)   15.0000  ( -VCC)  -15.0000  (N00317)     .5000 (N00348)    -.9983

(N00485) 25.94E-06 (X_U1.6) 125.8E-06 (X_U1.7)   -1.1085 (X_U1.8)   -1.1085

(X_U1.9)    0.0000 (X_U1.10)    -.6077                  (X_U1.11)   14.9600

(X_U1.12)   14.9600                   (X_U1.13)    -.5938
```

741 op amp-...

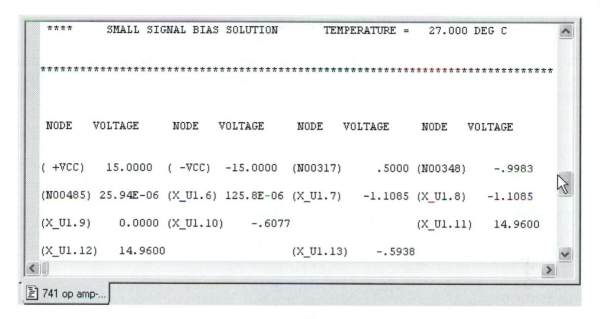

Figure 56 Amplifier simulation results using the 741 op amp.

```
VOLTAGE SOURCE CURRENTS
NAME          CURRENT

V_V3           1.667E-03
V_V1          -5.000E-05
V_V2          -1.667E-03
X_U1.vb        1.258E-09
X_U1.vc        1.500E-11
X_U1.ve        1.300E-11
X_U1.vlim     -2.203E-03
X_U1.vlp      -4.220E-11
X_U1.vln      -3.780E-11
```

741 op amp-...

Figure 57 Amplifier simulation results showing current flows.

5.3   *Using PSpice Subcircuit Models*

One of the powerful features of SPICE is the ability for the end user to develop new circuit element models either for devices not included in the libraries, or to idealize or simplify certain electrical behaviors. These new models are called subcircuits in SPICE. OrCAD Capture and PSpice facilitate the use of subcircuits by providing the ability for a user to create a new schematic part that is linked with a simulation model. A subcircuit, sometimes called a macromodel, is analogous to a procedure in a programming language. Subcircuits consist of a netlist that defines the function of the new part by interconnecting PSpice primitive elements like resistors, capacitors, inductors, sources, etc.

We will illustrate the usefulness of subcircuits and the ability to build new circuit models through an example in which we will modify the behavior of an existing model. We will not work through the exercise of developing an entire new schematic part from scratch since that exercise is beyond the scope of this book.

Example 9: Modify the existing μA741 op amp part so that its PSpice model exhibits the behavior of the ideal op amp. Save the new part so that it can be used in simulations of future simplified amplifier circuits. Use the new ideal op amp part to simulate the behavior of an inverting amplifier over a range of input voltages.

Solution 9: Begin by examining the netlist for the amplifier of Figure 55. Click on the menu item PSpice>View Netlist in the OrCAD Capture window for the schematic shown in Figure 55. You should see a window similar to the one in Figure 58. The first element in the netlist is the line containing X_U1, which represents the circuit connection to the element model for the μA741 op amp model. We refer to this line as the "calling statement"; much like procedure calls in programming languages, this statement "calls" the subcircuit and incorporates it into the original netlist. Five node connections are noted on this element statement beginning with a connection to node 0. The link to the subcircuit representing the op amp model is specified on the same line by the parameter "uA741."

To view the subcircuit that contains the circuit elements that model the behavior of the μA741 op amp, first select the op amp part in the schematic of Figure 55, then click on the menu item Edit>PSpice Model. A window similar to the fragment shown in Figure 59 will appear. The first seven lines of the file in Figure 59 are comments. Line 8 is the statement that links back to the "calling" netlist. The PSpice keyword .SUBCKT indicates that the following statements are a part of the *macromodel* for a circuit element. The subcircuit ends with the PSpice key word .ENDS at the bottom of the netlist. Node numbers in the .SUBCKT statement provide the links back to the calling netlist. These node numbers are only valid locally (i.e., within the subcircuit definition) and are connected to the original calling netlist by position. For example, node 1 in the subcircuit is connected to node 0 in the calling netlist; node 2 in the subcircuit is connected to node N00485 in the calling netlist; node 3 in the subcircuit is connected to node +VCC in the calling netlist; etc.

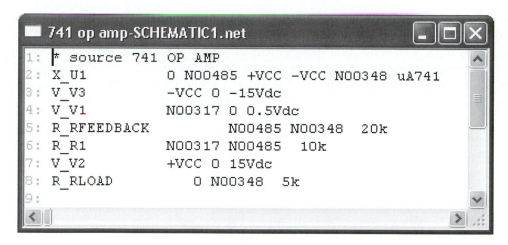

Figure 58 Netlist for the amplifier schematic of Figure 55.

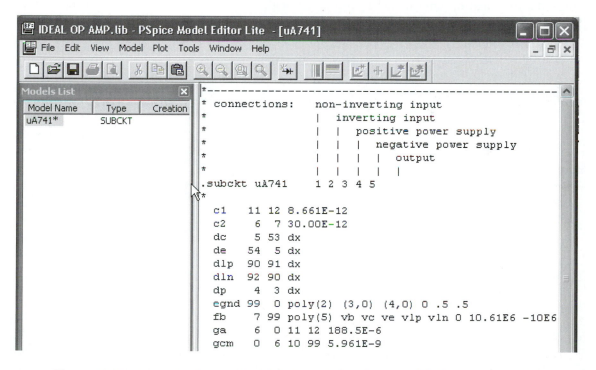

Figure 59 Portion of the PSpice Model Editor window for the μA741 op amp model.

Following the .SUBCKT statement, the remaining statements represent circuit elements that define the behavior of the subcircuit. We see that the model for the μA741 op amp contains a number of resistors, capacitors, and independent as well as dependent sources. You can easily see that the model is quite complex, accounting for the elaborate behavior of the physical μA741 op amp integrated circuits chip. We will develop a simplified version of this subcircuit that consists of the ideal op amp model discussed in Example 7 and shown in Figure 51.

We begin to create a new idealized op amp model by opening a new schematic window and creating an instance of the μA741 op amp from the eval.olb library using the Place Part button. Select the μA741 op amp part in the schematic capture window and click on the menu item Edit>PSpice Model. The Model Editor window of Figure 59 will appear. In the Model Editor window, choose the menu item Model>Copy From… which will produce a window titled Copy From. Click on the Browse button and navigate to the filename …OrcadLite\Capture\Library\Pspice\eval.lib. Scan down the list labeled From Model and choose the model μA741. Type in a name in the New Model field—we use the name IDEALIZED_OP_AMP—and click OK. Edit the model statements so that the resulting subcircuit model looks like the one in Figure 60. The subcircuit model typed into this window represents the ideal op amp model in Figure 51 with R_i = 500 MΩ, A = 10^8, and R_0 = 10 Ω. Remember that all of the node numbers/names in the subcircuit are local to the subcircuit alone except for the reference node 0. Specifying the node number 0 in the subcircuit provides a global connection to the reference node 0 for the overall circuit.

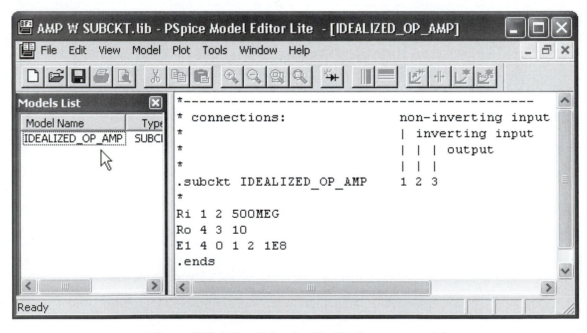

Figure 60 Subcircuit for the idealized op amp model.

Next we must modify the schematic symbol that represents the new subcircuit to reflect the fact that there are fewer pins and give the part a new name to distinguish it from the μA741 standard library part. To do this, close the Model Editor window and select the new part symbol in the schematic capture window. Click on the menu item Edit>Part to produce a graphical editor window for the part diagram. Choose and delete the graphics entities that are not needed for the idealized version of the new op amp model. The pin numbers may be eliminated by double-clicking on a pin and deleting the pin number in the Pin Properties window. To change the part name, double-click on the current name and type in the new name. Click the X button to close the graphic editor window and the Update Current button when it appears. The resultant part symbol should look like the one in Figure 61.

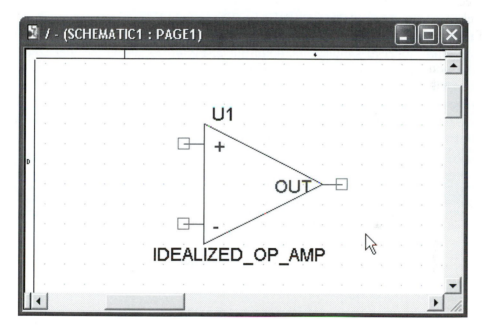

Figure 61 Part symbol for the idealized op amp.

We must take one more step before the new idealized op amp model is ready to be used in a PSpice simulation. Double-click on the new symbol in the schematic capture window to open the Property Editor. Change the Implementation field by entering the name IDEALIZED_OP_AMP which connects this instance of the symbol to the new subcircuit model. Next, modify the PSpice Template field so that it is the same as the one in Figure 62. In doing this edit you will remove the external connections to the +V and −V voltage supply pins which are no longer present in the new simplified subcircuit. Be sure to click Apply and close the Property Editor.

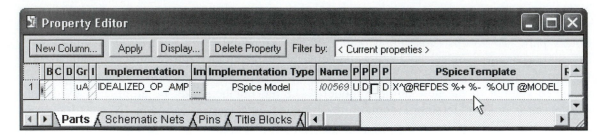

Figure 62 Adjusted PSpice Template parameter to remove %+V and %-V references.

The new simple op amp model called IDEALIZED_OP_AMP is now ready to be used in the simulation of a complete amplifier circuit. To illustrate its usefulness, we will employ the idealized op amp model to simulate the amplifier in Figure 52, using the idealized op amp in place of the voltage-dependent voltage source and associated resistors. The amplifier circuit based on the IDEALIZED_OP_AMP is shown in Figure 63. Simulation of this circuit produces the result shown in Figure 64 with the output voltage being -1.000. As expected, the overall voltage gain for this amplifier is -2. You should also examine the netlist for this simulation and recognize that the subcircuit is properly connected to the rest of the amplifier.

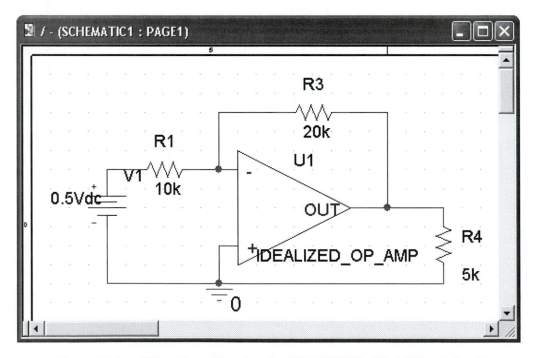

Figure 63 Amplifier schematic using the IDEALIZED_OP_AMP subcircuit.

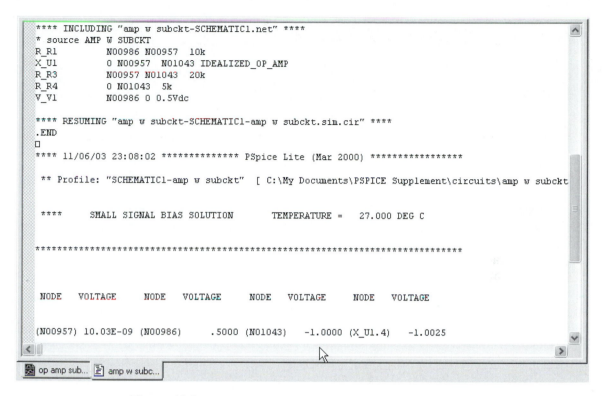

```
**** INCLUDING "amp w subckt-SCHEMATIC1.net" ****
* source AMP W SUBCKT
R_R1          N00986 N00957  10k
X_U1          0 N00957  N01043 IDEALIZED_OP_AMP
R_R3          N00957 N01043  20k
R_R4          0 N01043  5k
V_V1          N00986 0 0.5Vdc

**** RESUMING "amp w subckt-SCHEMATIC1-amp w subckt.sim.cir" ****
.END
□
**** 11/06/03 23:08:02 ************** PSpice Lite (Mar 2000) ******************

 ** Profile: "SCHEMATIC1-amp w subckt"  [ C:\My Documents\PSPICE Supplement\circuits\amp w subckt


 ****      SMALL SIGNAL BIAS SOLUTION          TEMPERATURE =   27.000 DEG C


 *********************************************************************************


 NODE    VOLTAGE      NODE    VOLTAGE      NODE    VOLTAGE      NODE    VOLTAGE


(N00957) 10.03E-09 (N00986)     .5000 (N01043)   -1.0000 (X_U1.4)   -1.0025
```

op amp sub...   amp w subc...

Figure 64 Output results for amplifier with idealized op amp.

Chapter 6 Time Domain Analysis

All electrical circuits have some type of time-dependent behavior that needs to be considered during the design cycle. Even simple electrical devices like the flashlight, physically consisting of a battery, a switch, and a resistor (the bulb), have a time-varying behavior. As the switch is moved from off to on, or vice versa, the response of the circuit is time-varying and should be analyzed in order to understand not only the circuit's performance, but also its reliability, viable lifetime, etc.

The term *time domain analysis* indicates that we will be determining the output variables as a function of time. Therefore, the independent variable in our calculations will be t, standing for the time elapsed since some arbitrary start-time. In SPICE calculations, time domain analysis is referred to as *transient analysis*. DC analyses performed earlier calculated values of the dependent variables as a function of voltage or current. One of the forms of output in a DC analysis is a plot of an output variable vs. voltage (or current). When performing transient analyses, we will generate plots of output variables vs. time. These results are similar to what we might generally see displayed on a typical oscilloscope.

Previous chapters have covered the DC behavior of circuits where the only circuit elements considered were DC sources and resistors. Any capacitances that might have been present in a physical circuit were modeled as open circuits while inductors were modeled as short circuits. In this chapter, time-varying sources will be considered and all capacitance and inductance will be included in the analysis. In many circuit analysis books, the response of a circuit driven by a time-varying source is referred to as the *forced response* of the circuit—the voltage source is the forcing function. Following the lead of typical circuit analysis texts, we will begin our discussion of time-varying circuits by analyzing circuits that contain no sources. For these circuits we will be seeking what is referred to as the *natural response* of the circuit. Initial conditions will be specified for an arbitrary time called t = 0. Typically, some event will occur at time t = 0 that will change the state or the configuration of a circuit and we are interested in knowing the circuit's voltages and currents for t > 0.

6.1   *Source-Free RL Circuits*

Source-free circuits must have an initial voltage on a capacitor or an initial current flowing in an inductor in order for a time-dependent response to occur. The first step in performing an analysis on a source-free circuit is to determine the initial conditions. Once the initial conditions are known, the circuit can be drawn, the element parameters entered (including initial conditions), and the simulation performed.

Example 10: Determine the natural response of a simple RL circuit with R in parallel to L. Set R = 1 kΩ and L = 10 mH with an initial current of 10 mA flowing in the inductor at t = 0. Plot the loop current using Probe and determine the time at which the current has reached ½ of its original value.

Solution 10: The schematic for the circuit described above is shown in Figure 65. For these simple circuits it is important to remember to include the ground node in order to avoid perplexing error messages.

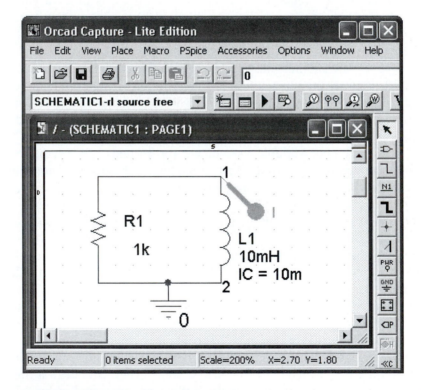

Figure 65 Source-free RL schematic with initial conditions and current sensing.

A current marker is used in the schematic capture window to indicate that the current entering into the top node of the inductor is to be plotted. The initial current in the inductor is specified by opening the property editor for the inductor and entering 10m in the field labeled IC. IC stands for initial condition. Amperes is assumed to be the units since only initial currents can be specified for inductors. To display the initial condition in the schematic as shown above, with the

IC field selected, click on the Display… button in the property editor and select the radio button Name and Value. Click OK, click Apply, and then close the property editor window.

Details of the transient analysis are set up by opening the Simulation Settings window and selecting the PSpice>New Simulation Profile menu item. Choose the analysis type to be Time Domain (Transient). Since the initial conditions for the circuit are specified as part of the inductor parameters, we check the box "Skip the initial transient bias point calculations (SKIPBP)." This saves simulation time by omitting unnecessary attempts at re-calculating the state of the circuit at t = 0. Finally, we must set the "Run to time" or the TSTOP parameter. The circuit will be analyzed from t = 0 until t = TSTOP. This means that TSTOP should be selected so that the circuit is analyzed over the full range of time values of interest for this problem. From our knowledge of circuit analysis we know that this circuit should exhibit a current flowing in the inductor that starts at 10 mA and decays exponentially toward zero. The exponential decay time is dependent on the time constant $\tau = L/R$. Once t = 5τ, the exponential becomes essentially zero and stays that way ad inifinitum. Therefore, the value of TSTOP should be set to 50 μs for this problem in order to observe the full range of interesting behavior of the output current. The resulting Simulation Settings window should look like the one in Figure 66.

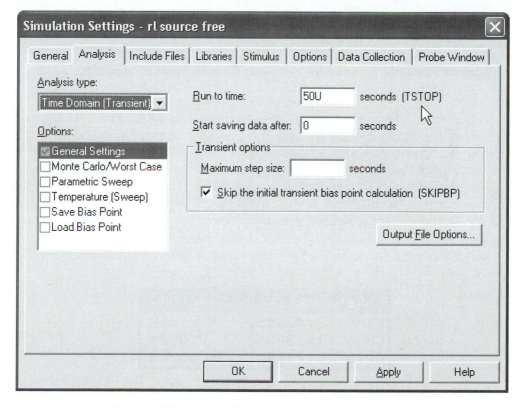

Figure 66 Simulation Settings for a transient analysis.

After clicking Run, PSpice will produce the Probe plot seen in Figure 67. As expected, the output current decays exponentially to approximately zero near t = 50 μs. To determine the time at which the current reaches 50% of its initial value, click on the Toggle Cursor button.

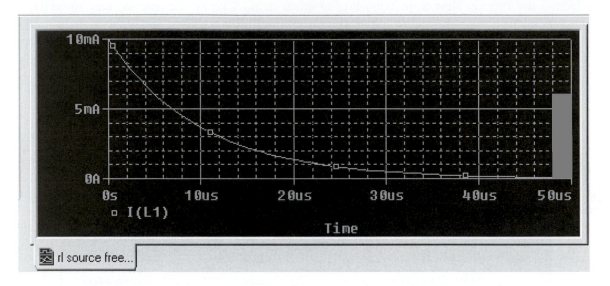

Figure 67 Plot of the time-varying current in the inductor of the simple RL circuit.

A window titled Probe Cursor appears in the lower right corner of the Probe window similar to the one in Figure 68. It displays the position of the primary cursor in row C1. Row C2 can be used to display the coordinates for a second cursor when a second variable is plotted. Cursors follow the data plot and are moved from point to point either by clicking on the right or left arrow keys or by dragging the cursor using the left mouse button. The row labeled "dif" in the Probe Cursor window display the difference in the current cursor position and the last cursor location marked as a reference point. A reference point is marked by placing the mouse pointer over a portion of the plot window and clicking the right mouse button. Data in Figure 68 shows that the current in the circuit is reduced to 50% of its initial value, or 5 mA, at approximately 4.96s.

```
Probe Cursor
C1 =   7.0238u,    4.9564m
C2 =  10.000n,     9.990m
dif=   7.0138u,   -5.0336m
```

Figure 68 Probe Cursor window showing the point at which the current is approximately 5 mA.

In this section we have performed a simple transient analysis for a source-free circuit and have exercised the ability to use the plot program to visualize the circuit response as well as discern specific response values from the plotted output. In the next section we will learn to use other types of transient response components and simulation tools.

6.2   Source-Free RC Circuits

A simple, yet important circuit, is one consisting of a charged capacitor discharging through a resistor. This circuit forms the basis for many different component behaviors seen in electronic devices such as diodes, transistors, etc. The basic behavior of this source free circuit can be modeled in much the same way as the RL circuit above. However, for the purposes of illustration, we will introduce a new transient element, the switch, and use it to connect the capacitor to the discharge path at an arbitrary time.

Example 11: Configure a 5 kΩ resistor in series with a switch, in series with a 10 μF capacitance. Assume that the capacitor is initially charged to 5 V and is to be placed in parallel with the resistor at time t = 10 ms. The configuration should model the situation shown in Figure 69. Determine the voltage across the capacitor and the current through the resistor as a function of time.

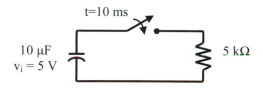

Figure 69 RC circuit with switch closing at t = 10 ms.

Solution 11: A schematic model for this circuit is shown in Figure 70. The capacitor is a standard part from the ANALOG library as is the resistor. Using the property editor, the initial voltage across the capacitor is specified to be 5 V. As in the previous example, the IC parameter in the property editor is modified and set up to display in the schematic window. The polarity of the initial capacitor voltage is determined by the position of the capacitor as it is placed in the schematic capture window. Initially, the positive node is on the left of the capacitor and it changes as the part is rotated to position in the figure. We have rotated the capacitor three times to get the positive node at the top of the capacitor.

A model for the switch is found in the EVAL library in the form of a part named SW_tClose. Using the property editor window for the SW_tCLose part shown in Figure 71, we have set the switch to close at t = 10 ms in the parameter field labeled TCLOSE. Transition time for the switch, i.e., the time it takes the switch to close, is specified in the TTRAN field as 10 ps. This is a very fast

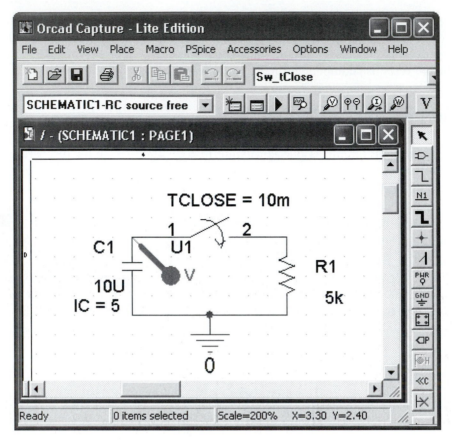

Figure 70 RC circuit with switch closing to drain capacitor. Switch closes at t = 10 ms.

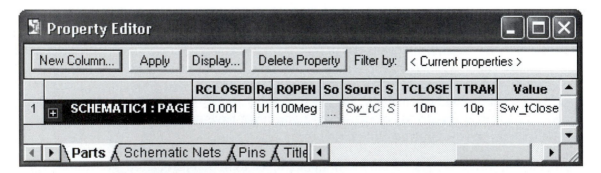

Figure 71 Property editor for the PSpice switch part.

switch closure especially when considered in relation to other circuit time constants. The open-circuit and closed-circuit resistance of the switch are specified in fields ROPEN and RCLOSED, respectively. Values for these parameters are set so that the simulation part closely models an ideal switch.

Results from the simulation of the RC circuit of Figure 70 showing the voltage across the capacitor are given in Figure 72. Notice that the voltage stays at five volts for 10 ms and then begins to drop exponentially toward zero. The voltage curve reaches zero at approximately 250 ms, which is roughly 5τ since $\tau = RC = 10 \ \mu F \times 5 \ k\Omega = 50$ ms.

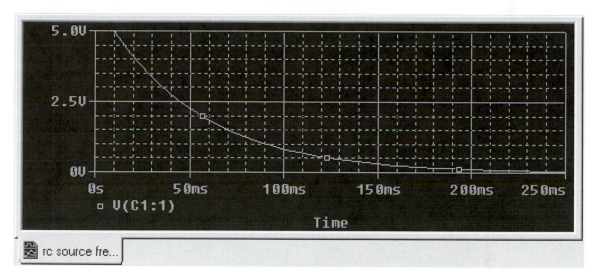

Figure 72 Plotted results for the RC circuit with the switch closing at t = 10 ms.

To obtain a plot of the current in the resistor we need to modify the output plot requested in Probe. First, in the Probe window click on the menu item Trace>Delete All Traces. Next, click the menu item Trace>Add Trace… and the window shown in Figure 73 will appear. Choose the Simulation Output Variable I(R1) to plot the current into the positive terminal of the resistor. Note that the resistor was rotated three times from its original position in the schematic window in order to place the node designated as the resistor's positive node in the topside position. Although a resistor has no inherent polarity, PSpice uses a convention to designate one of the terminals as the positive node and then reports I(R) as the current into the positive node. A plot of the current into the resistor is shown in Figure 74. In order to display the maximum range of current in the plot window, modify Probe's default settings by clicking on the menu item Plot>Axis Settings…. Select the Y Axis tab at the top of the window and click on the User Defined radio button in the Data Range area. Fill in the values so that the Y-axis ranges from 0A to 1 mA. For this situation, these manual settings spread the Y-axis better than the default Auto

Range settings allowing us to better visualize the exponential decrease in current that occurs when the capacitor is discharged through the resistor. We should also recognize the 10 ms delay that occurs before the discharge current begins to flow. This of course is due to the fact that the switch is set to close 10 ms at t = 0.

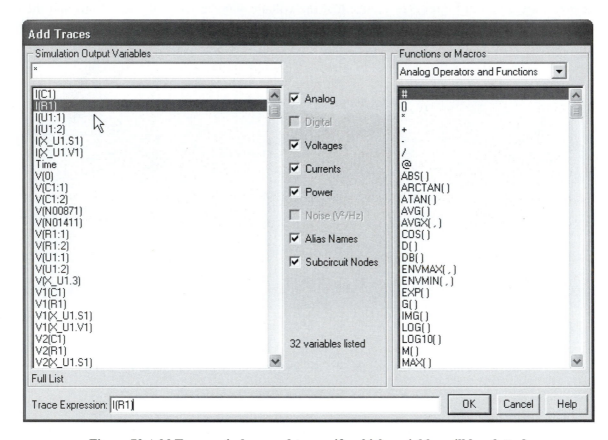

Figure 73 Add Traces window used to specify which variables will be plotted.

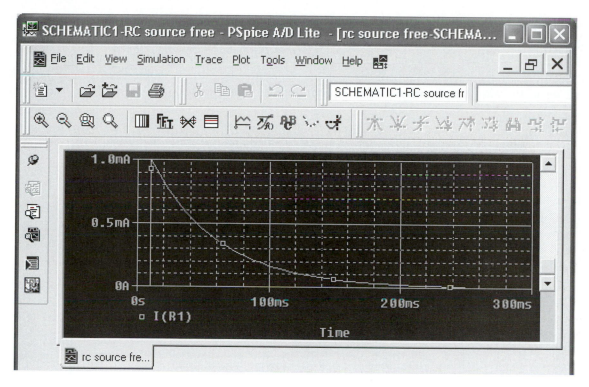

Figure 74 Probe window with a plot of the resistor current for the RC circuit.

6.3   Source-Free RLC Circuits

Most practical circuits include all three main electrical components: resistors, inductors, and capacitors. Many times these circuit elements are not all included intentionally; rather some are parasitic elements that become part of the circuit because of the way in which it is fabricated. To examine the effects of these types of circuits, we will analyze a source-free RLC circuit and determine one type of response that can occur.

Example 12: Simulate the behavior of a circuit consisting of a 5 Ω resistor in series with a 2 mF capacitor in series with a 5H inductor. The capacitor is initially charged to 10 V and the inductor has 1 A flowing in it at t = 0. (Notice that these component values are not entirely practical, but are useful to illustrate basic calculations and principles). Determine the voltage and current through the resistor.

Solution 12: Transient analysis of the RLC source-free circuit is performed in much the same way as earlier time domain simulations. A schematic for the circuit is shown in Figure 75. Initial conditions have been entered using the property editor windows for the capacitor and the inductor. Voltage and current probes are used to monitor the variable values across and through the resistor. The simulation settings are configured to run the simulation from t = 0 to t = 5 seconds.

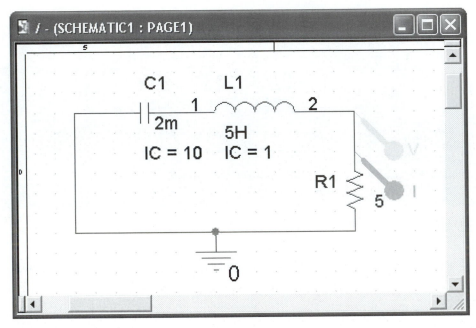

Figure 75 Source-free RLC circuit with initial conditions.

The results of the simulation are shown in Figure 76. Both the voltage and the current are shown on the same plot. The waveform in the figure is underdamped with the voltage starting off at 5 V while the current is 1 A. Both values eventually fall to zero as expected in a source-free circuit.

Confirmation of the general shape of the waveforms can be obtained through two simple calculations:

$$\alpha = \frac{R}{L} = \frac{5}{5} = 1$$

$$\omega_0 = \frac{1}{\sqrt{LC}} = \frac{1}{\sqrt{.01}} = 10$$

Since $\omega_0 > \alpha$, the response of the circuit is underdamped and the waveforms calculated by PSpice are reasonable.

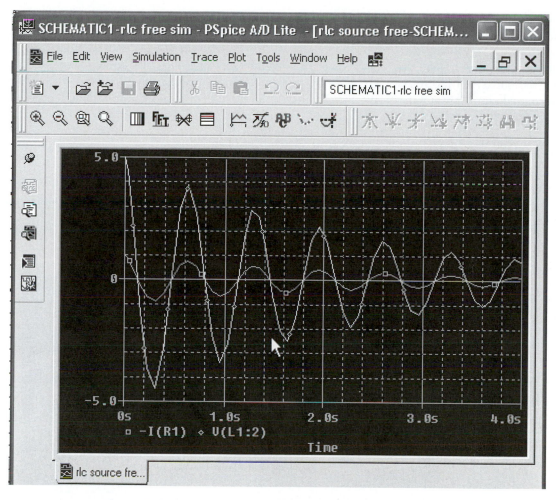

Figure 76 Simulation results for the RLC source-free network.

The waveforms in Figure 76 are not as smoothly shaped as the waveforms that would be observed in an actual circuit. Because of the default values in the simulations settings window, there was not a sufficient number of data points collected during the simulation to provide the details of the rounded peaks and valleys of the actual waveform. To obtain a better plot of the actual waveform, we make an adjustment to one of the parameters in the Simulation Settings window. The number of data points collected during a transient analysis simulation run is determined by the simulation step size. PSpice automatically calculates the simulation step size based on the sizes of the capacitors, inductors and resistors in the circuit. Set the step size to a smaller value by opening the Simulation Setting window (click PSpice>Edit Simulation Profile) and input a value into the field labeled Maximum Step Size. Entering 0.001 seconds as the maximum step size produces the plot shown in Figure 77 when the simulation is re-run. A total of 4001 data points are collected from this simulation, producing a much smoother and more realistic plot of the time-dependent behavior of the circuit.

It is worth noting the phase relation between the voltage and the current. The voltage is in phase with the current in the resistor as it is expected to be since the relationship between the two is determined simply by Ohm's Law. The reader should use Probe to plot the voltage and current through both the inductor and the capacitor and examine the phase relationship between them.

Notice that the current variable plotted in both Figure 76 and Figure 77 is designated as –I(R1). The negative of the variable is reported because of the way in which the resistor was placed in the schematic—i.e., the positive node of the resistor was connected to the ground terminal when the part was rotated into the vertical position. Placing the Current Marker at the top node of the resistor causes the current into that node to be monitored.

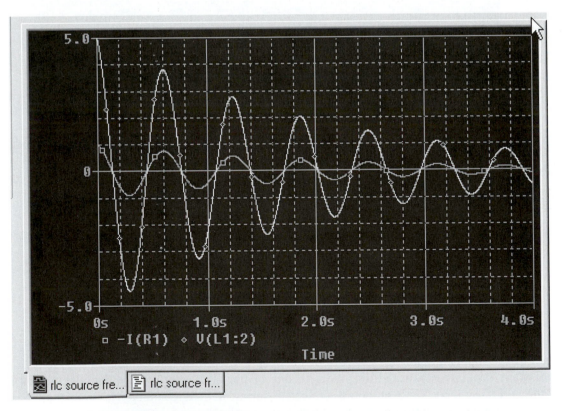

Figure 77 Simulation results for the RLC circuit with a small time step size.

6.4 Time-Varying Sources

Most practical circuits have some form of time-varying source in the circuit. In simple cases time-varying sources are modeled using switches. More complex time-varying sources require models that follow certain mathematical behavior. In this section we will explore the different types of sources that can be modeled with PSpice's built-in time-varying source models. Time-varying source models are contained in the SOURCE parts library. For the following parts descriptions, seconds are the units for all variables that are a function of time; volts are the units for all variables that describe voltage; and amperes are the units for all current parameters.

Voltage source VPULSE is one of the simplest time-varying voltage supply models. This voltage source delivers a pulse waveform which is defined by a set of parameters that control transition times and voltage levels. A description of the behavior of the VPULSE source is shown in Figure 78. In order to use the source, place it in a circuit like any other part and then fill in the controlling parameters. Parameter values may be supplied by clicking on the parameter name and typing a value. Alternately, double-click on the source symbol, which will open the property editor, then enter values in the appropriate fields.

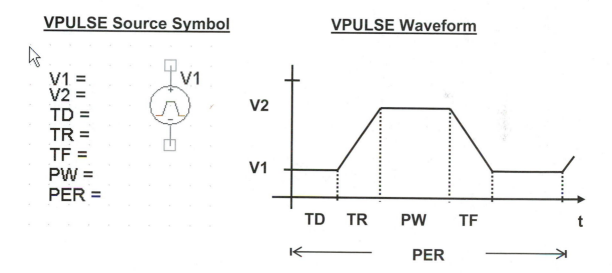

Figure 78 Definitions for time-varying source VPULSE.

Voltage source VSIN produces a time-varying sinusoidal waveform as shown in Figure 79. Variable VOFF defines the initial offset voltage for the waveform. The waveform amplitude is specified by the variable VAMPL with the actual maximum voltage being VOFF + VAMPL. TD is the time delay before the sine wave begins oscillating at a frequency of FREQ. DF is the time constant for the damping factor which causes the amplitude of the sine wave to decay exponentially to the value of VOFF. Each of these parameters is input into the model either by double-clicking on the parameter name in the schematic capture window or by opening the

Simulation Settings window and entering parameters in the appropriate columns. Parameters may be set to display (or not) in the schematic by using the Display button in the property editor window for the VSIN part. Both TD and DF have default values of zero. The default value for FREQ is 1/TSTOP Hz, where TSTOP is the simulation stop time as specified in the Simulation Settings window.

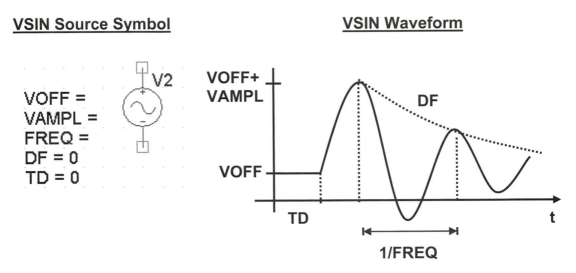

Figure 79 Definitions for the time-varying source VSIN.

The part named VEXP is a time-varying voltage source similar to the other time-varying sources. It produces an exponentially rising and exponentially falling voltage waveform as shown in Figure 80. Parameters V1 and V2 set the offset and peak voltages, respectively, for the VEXP waveform. TD1 is the time delay before the start of the rise of the waveform. TC1 is the exponential time constant for the rising portion of the waveform. The waveform rises exponentially until time t = TD1 + TD2 after which the voltage begins to fall exponentially with a time constant TC2. By adjusting TC1 to be very small as compared to other circuit time constants, the VEXP can be set up to produce what is essentially only an exponentially falling waveform. Likewise, the parameters can be set so that the waveform rises exponentially and essentially remains at V2 for the remainder of the simulation period. The default value for TD1 is zero, while the default value for TD2 is TD1 plus one simulation time step TSTEP as set in the Simulation Setting window.

VPWL is a very versatile voltage source that produces a piecewise linear time-varying waveform. A VPWL voltage source produces the waveform shown in Figure 81. Voltages transition linearly with respect to time from point (T1,V1), to (T2,V2), to (T3,V3), etc. PSpice allows for eight pairs of end points for the waveform. Users may compose a variety of waveforms approximating a numbers of different real-world situations with this source, including switching events that occur in digital circuits.

VEXP Source Symbol

VEXP Waveform

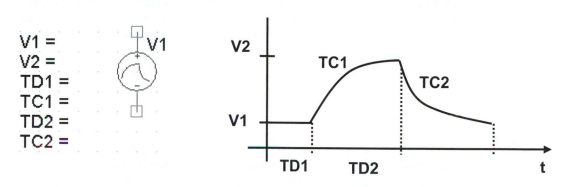

V1 =
V2 =
TD1 =
TC1 =
TD2 =
TC2 =

Figure 80 Definitions for the time-varying source VEXP.

VPWL Source Symbol

VPWL Waveform

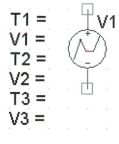

T1 =
V1 =
T2 =
V2 =
T3 =
V3 =

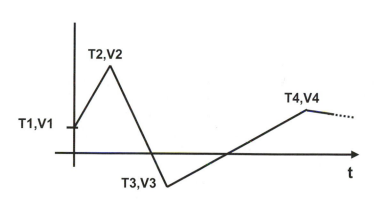

Figure 81 Definitions for the time-varying source VPWL.

For each of the time-varying voltage sources described above, there is a corresponding current source that has the same set of parameters and produces the same format of waveform. The corresponding part names are IPULSE, ISIN, IEXP, and IPWL.

6.5   *Circuits with Time-Varying Sources*

The step response of a circuit is its behavior when the voltage or current excitation is applied in the form of a step function. The ideal step function has an infinitely small rise time. Because of the way in which PSpice performs its calculations, we are limited to producing a driving source that has a finite slope for its rise or fall time.

Example 13: Determine the step response of a simple RC series circuit which has an initial voltage across the capacitor of zero volts.

Solution 13: A schematic for the simple RC circuit is shown in Figure 82. Source VPULSE is used to provide the step function input which changes from zero to five volts. The rise time of the input pulse is set to be 1 nanosecond, or 10^{-9} seconds in order to approximate a pulse that has a zero rise time. (As mentioned earlier, PSpice cannot generate waveforms with exactly zero rise times). TD is set to zero so that VPULSE begins rising immediately at t = 0. VPULSE parameters are set using the Property Editor window and each parameter is set to display in the schematic capture window.

A Time Domain (Transient) analysis is specified in the Simulation Settings window along with ensuring that the SKIPBP check box is left unchecked. SKIPBP is left unchecked to enable an initial bias point analysis of the circuit to be performed. This bias point analysis will establish the initial conditions for the capacitor, which will be $V_c = 0$ since no initial voltage is specified for the capacitor. The time constant τ for the circuit is 1 millisecond, which is obtained by multiplying the value of the resistor by the value of the capacitor. TSTOP is set to 10 milliseconds in the Simulation Settings window in order to allow more than a full 5τ of the response waveform to be simulated. The PW parameter for VPULSE is set to 100 milliseconds so that the voltage input waveform stays at 5 V throughout and beyond the interval of interest for this simulation.

The simulation results for the capacitor voltage are shown in Figure 83. As expected, the voltage waveform follows the equation:

$$V_C = 5(1 - \varepsilon^{-\frac{t}{.001}}),$$

effectively reaching its final voltage of 5 V volts after about 5 milliseconds.

It is interesting to examine the current into the capacitor at the same time we view the capacitor voltage. This can be done without re-running the simulation but rather by making adjustments to the parameters of the plotting program Probe.

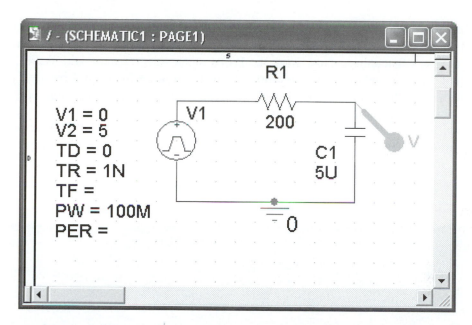

Figure 82 Schematic for the series RC circuit driven by VPULSE.

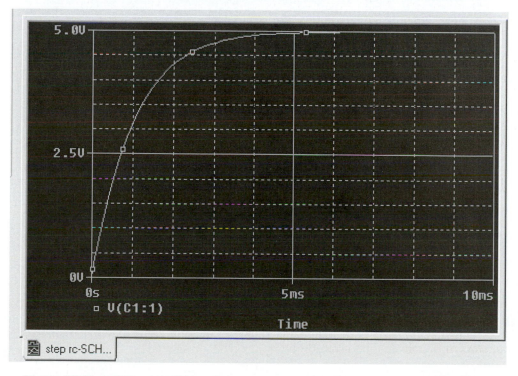

Figure 83 Plot of the capacitor voltage response to the five volt step function input.

From the Probe menu select Plot>Add Y Axis and you will see a second Y-axis appear to the left of the original axis. This axis will show the scale for the first variable plotted—in this case, the voltage across the capacitance. Now from the Probe menu select Trace>Add Trace... and click on the current variable I(R1). The waveforms shown in Figure 84 should be displayed in the Probe window. Each of the Y-axes will have independent scaling allowing you to readily compare events occurring in the two different waveforms. From the figure we can see that as the voltage exponentially rises to its maximum of 5 V, the current exponentially decreases toward 0 mA. This reflects the fact that the capacitor is nearly fully charged to 5 V after about 5 milliseconds and requires a decreasing amount of charge to be transferred to it from the voltage source as time goes on.

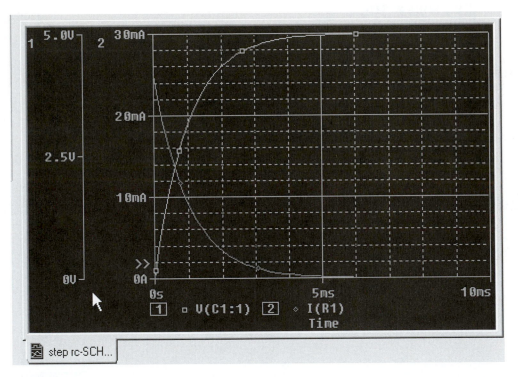

Figure 84 Dual plot of the voltage and current response for the RC circuit driven by VPULSE.

A slightly more complex example of the use of a time-varying source is presented in the next.

Example 14: The RLC circuit shown in Figure 85 is to be analyzed using PSpice. Determine the behavior of the current flowing in the inductor and the voltage across the capacitor.

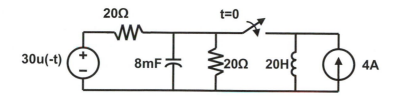

Figure 85 Determine the current in the inductor for the RLC circuit.

Solution 14:

Solution 14:   At time t < 0 the switch in Figure 85 is open, allowing the capacitor to be charged to 15 V, and the inductor to carry an initial current of 4 A. A schematic for the circuit is shown in Figure 86. In this schematic, the VPULSE source emulates the behavior of the step function voltage source 30u(-t) by having an initial voltage of 30 V for t < 0 and changing to 0 V at t = 0. The rise time for VPULSE is set to 1 nanosecond to model a very sharply falling pulse as defined by the u(-t) function. To insure that the initial voltage on the capacitor is properly set, we use the property editor for the capacitor and set IC = 15 V. Similarly, the property editor is used to set the initial current in the inductor to 4 A.

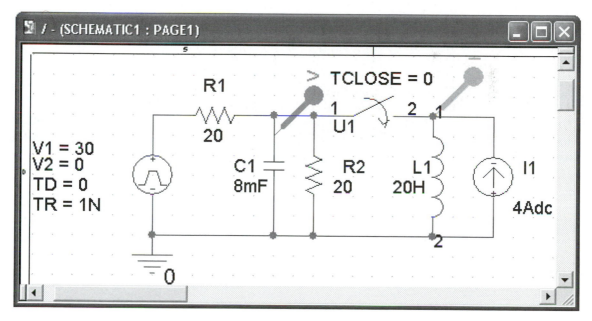

Figure 86 RLC circuit with step input and switched current source.

After making some rough estimates of the expected time constant for the system, the TSTOP time is set to 5 seconds in the Simulation Settings window. An alternate method for choosing an appropriate value for TSTOP is to start by using the default setting, examining the simulation results, then recognizing that TSTOP must be adjusted in order to view the entire relevant portion of the output waveform.

The output waveform for the simulation of the RLC circuit is shown in Figure 87. Plots of both the current in L1 and the voltage across C1 are shown. Independent Y-axes are set up by selecting the menu item Plot>Add Y Axis in the Probe menu. Next we scale the new axes by going to the Axis Settings window, which is reached by using the Plot>Axis Settings... menu selection. Click on the Y Axis tab and set the range for each of the Y-axes so that the full range of the waveform is displayed. First, choose the Y-Axis number from the pull down menu on the right. Next click the radio button for User Defined range and set the values in the areas below so that the full range of the waveform is displayed. Setting Y-axis #1 current to range from 4.0 to 4.06 A and Y-axis #2 voltage to go from -1 to 15 V displays the full range of the output waveforms.

Calculating the output voltage and current for this circuit produces the following results:

$$I_L = 4 + 0.063(\varepsilon^{\frac{-t}{0.0001}} - \varepsilon^{\frac{-t}{.334}})$$

and

$$V_C = 15(\varepsilon^{\frac{-t}{.0222}} - \varepsilon^{\frac{-t}{.0222}})$$

These results correspond very well with the results produced by the PSpice simulation. As shown in Figure 87, PSpice predicts that the current waveform rises from 0 A to nearly 4.06 A, then falls exponentially toward 0 A. Similarly, the voltage across the capacitor falls exponentially from 15 V toward -1 V then rises exponentially toward zero volts as the capacitor discharges through the resistors.

Transient analysis is an important part of most circuit design processes. The time-dependent behavior of a circuit must be examined in order to ensure that the network will perform properly in all of its operating environments. PSpice can be used to support the analysis of complex networks and provide accurate descriptions of voltages, currents, power dissipation, etc. as they change with time.

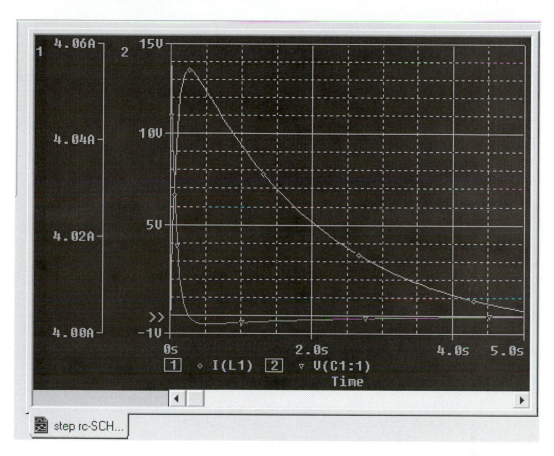

Figure 87 Output waveforms for the RLC circuit of Figure 86.

Chapter 7 Frequency Domain Analysis

Circuits have different behaviors when stimuli are applied at different frequencies. These response characteristics are significant in many different applications, e.g., in the tuning circuits of communication systems. It is important to be able to characterize these different behaviors in order to know that a circuit will perform properly in its designated frequency range. PSpice can perform analyses over ranges of frequencies and thus calculate values of voltages and currents as a function of frequency. This is generally referred to as frequency domain analysis and is also referred to sometimes as an .AC analysis or AC Sweep analysis, in the jargon of SPICE users.

7.1 Frequency Response

The frequency domain analysis performed by PSpice assumes that all transient effects in a circuit have come to a steady-state and no further time-varying changes will take place. All sources are assumed to be sinusoidally varying with constant amplitude throughout the analysis period. PSpice can calculate the amplitude and phase angle response at each specified frequency analysis point. To relate the PSpice calculations to the physical world, we note that a transient analysis produces an output like that seen on an oscilloscope set to view time-varying voltages or currents. Frequency Domain analysis produces an output that is similar to what you would see on an instrument known as a spectrum analyzer. We will demonstrate the frequency analysis capabilities of PSpice by examining the behavior of several filter circuits similar to those that might be used in communications and control circuits.

Example 15: Analyze the simple filter circuit shown in the schematic of Figure 88. The circuit consists of passive elements R1, R2, C1, and C2 that act as a filtering network. The source V1 is a steady-state sinusoidal signal generator with an input waveform having a magnitude of 1 V. Determine the frequency response of the circuit plotting both the magnitude and phase of the voltage across C2.

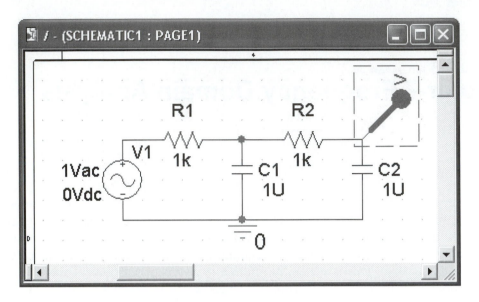

Figure 88 Schematic for a simple filter circuit.

Solution 15:   The schematic of Figure 88 is produced in the same way as previous circuit schematics except that the source V1 is based on the part called VAC from the SOURCE library. VAC generates a sinusoidal steady-state signal with a magnitude that has been set to 1 V and a phase angle set to zero. Each of these parameters can be set by either clicking the parameter on the screen or by using the Property Editor. A Voltage Marker is used to indicate that the voltage across C2 is to be plotted by Probe.

Menu item PSpice>New Simulation Profile is used to specify that a frequency response analysis is to be performed. The simulation settings shown in Figure 89 are set up to perform a frequency domain analysis sweeping the frequency from 1 Hz to 1000 Hz. Two hundred points will be plotted over the sweep interval. The independent axis (horizontal) will be divided linearly with data points being equally spaced across the axis. No additional settings are required for this simple analysis.

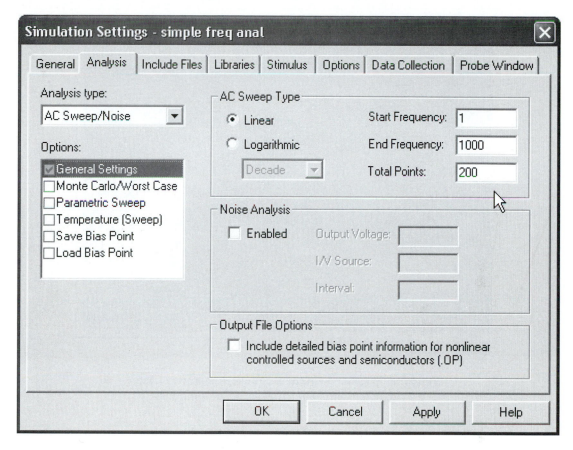

Figure 89 Simulation Settings window for a frequency response analysis.

A plot of the magnitude of the voltage across the C2 capacitor is shown in Figure 90. We can see that the circuit acts as a lowpass filter because the magnitude of the output signal is larger at lower frequencies while the signal strength provided by V1 is considerably attenuated at higher frequencies. Using the cursor we can locate various points of interest along the output waveform, most importantly the point where the output signal reaches $0.707 \times V1_{max}$. Usually referred to as the *cutoff frequency*, f_0, this point occurs at approximately 59.6 Hz for this filter.

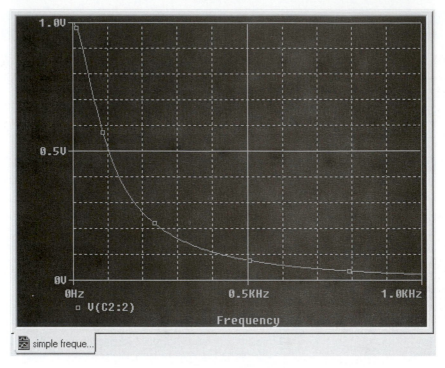

**Figure 90 Magnitude of the frequency response of the
C2 capacitor voltage for the circuit in Figure 88.**

Another important aspect of the output of this simulation is the relative value of the phase of the output. Input V1 is set to have a phase angle of zero. The phase of the output voltage waveform is compared to that of the input V1. It is sometimes important to view the phase of the output in relationship to the input. To do this, we could plot both values on the same axis. As an alternative, we will plot both outputs on individual axis, while using the same range of frequencies. With the plot shown in Figure 90 visible in the Probe window, execute the menu command Plot>Add Plot to Window, which will add a new set of axes to the plotting window. Next we must specify which circuit variable will be plotted by clicking on the menu item Trace>Add Trace… to open the Add Trace window shown in Figure 91. Choose the voltage V(C2:2) as the output variable by clicking on the line with this variable name in the left column. V(C2:2) is actually the voltage magnitude that is output in the original plot as specified by the voltage marker in the schematic. Now edit the line at the bottom of the window labeled Trace Expression so that it reads VP(C2:2) as shown in Figure 91. This notation indicates that the value of the phase angle of the voltage at node 2 of capacitor C2 is to be plotted. Once you click OK, the plot shown in Figure 92 appears. From the plots we see that over the range of frequencies from 1 Hz to 1 kHz, the magnitude of the capacitor voltage varies from 1 V to nearly 0 V while the phase angle of the voltage varies from 0° to nearly -154°.

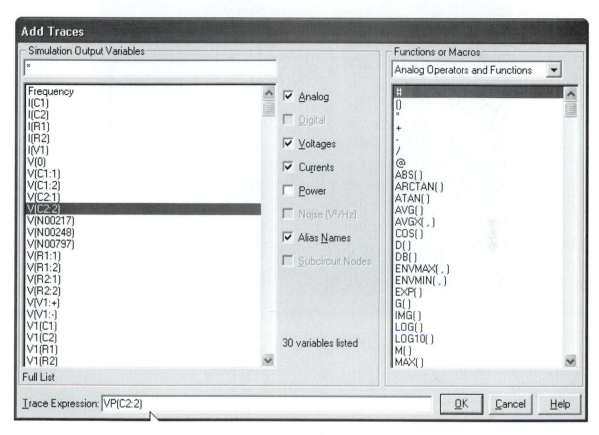

Figure 91 Plot of the voltage phase angle is added to the plot with the Add Traces window.

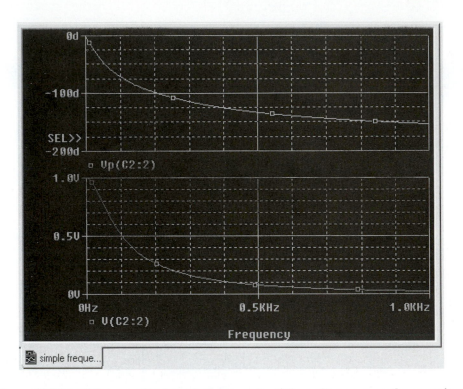

Figure 92 Plot of the magnitude and phase angle of the voltage across the capacitor.

Example 16: Use PSpice to plot the magnitude and phase of the voltages and currents for the circuit in the schematic of Figure 93. Use a logarithmic scale for plotting the frequency on the X-axis and determine the resonant and half-power (corner) frequencies for this bandpass filter.

Solution 16: The schematic of Figure 93 is developed much the same as the previous example. Source V1 is based on the sinusoidal steady-state source VAC with a magnitude of 1 V and a phase angle of 0°. The source values are chosen so that the simulation results effectively produce a transfer function for the magnitude of the output. To obtain the response of the circuit to signals of various magnitudes, we simply adjust the output waveform by multiplying the data points by the magnitude of the new source signal.

Simulation is set up by editing the parameters in the Simulations Settings window as shown in Figure 94. First the Analysis Type is set to AC Sweep/Noise in order to perform a frequency response analysis. Collection and plotting of the data points using a logarithmic scale for the X-axis is accomplished by clicking the radio button labeled Logarithmic and selecting either

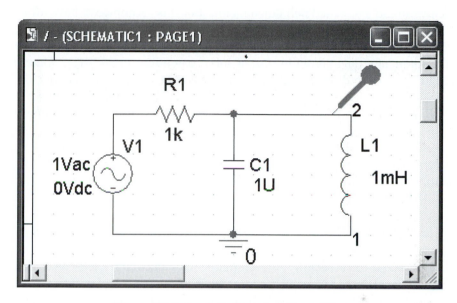

Figure 93 Schematic for a bandpass filter.

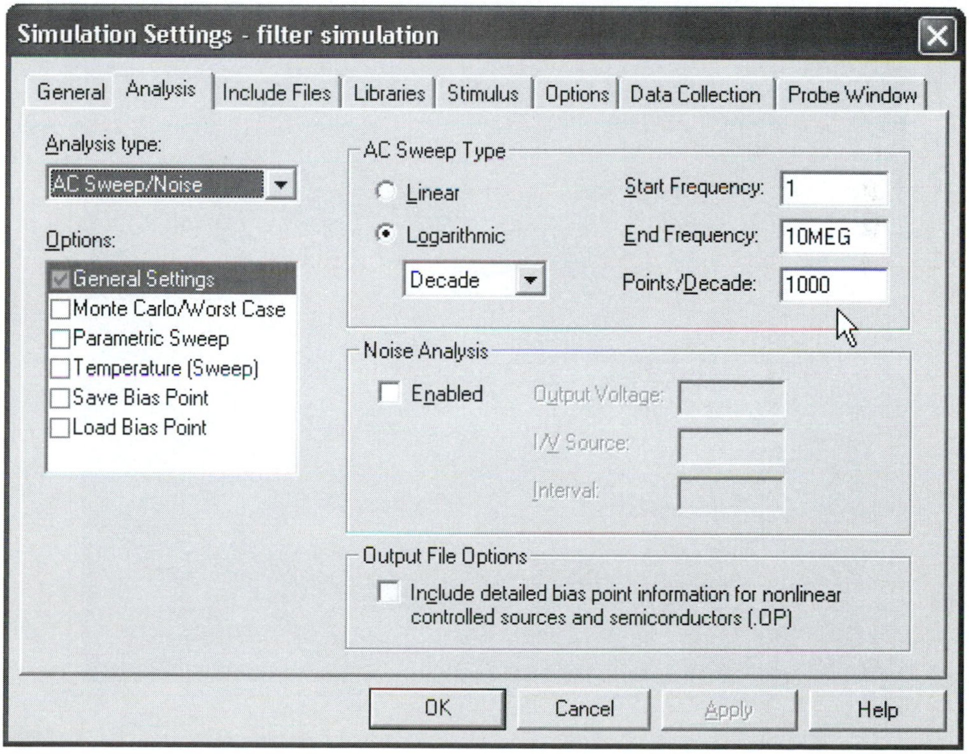

Figure 94 AC Sweep with output set to sweep over six decades and display logarithmically.

Decade or Octave in the associated pull down menu. For this example we have arbitrarily chosen Decade. If the behavior of the circuit is relatively unknown, a broad range of input frequencies can be used to get a better feel for the type of response that will be generated. For this simulation, the frequency range is set to start at 1 Hz, extending to 10 MHz. One thousand data points are set to be collected per decade of frequency. We choose the number of points to be collected somewhat arbitrarily, but we must be careful with this value. Choosing a value too low will not produce enough information in the output waveform requiring further simulations, while selecting a value that is too high results in lengthy simulation times. The value of 1000 points per decade produces over 6000 data points. Because of the small number of elements in the example circuit, 6000 data points will not take very long to generate. Larger circuits will of course require longer simulation times and will necessitate good judgment in selecting simulation settings parameters.

Initial simulation results for this circuit are shown in Figure 95. The general characteristics of the circuit behavior can be seen in this figure. The input is attenuated to nearly zero over much of the range of the plot except for a narrow band of frequencies between 1 KHz and 10 KHz. These frequencies are referred to as the pass band. Specific frequencies of interest in the pass band include the resonant frequency f_0 at which the output is a maximum and the two frequencies, f_1 and f_2, at which the output is $0.707 \times V_{max}$. In order to get a better estimate of the interesting frequencies, we perform a second simulation over a narrower range of frequencies and generating

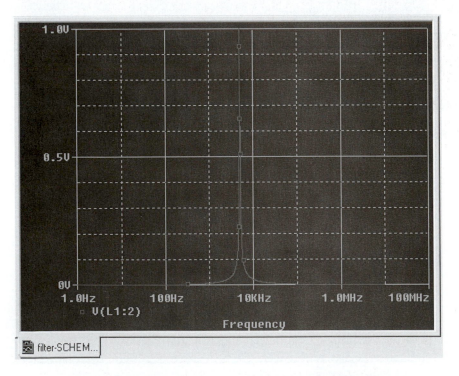

Figure 95 Initial simulation results for the bandpass filter.

a larger number of data points. Increasing the number of data points allows for a more accurate estimate of the frequencies f_0, f_1 and f_2.

Results from a second simulation with the frequencies ranging from 4 to 6 KHz are shown in Figure 96. Although the logarithmic frequency scale worked well when searching for the general overall shape of the response curve, this simulation was set up with a linear frequency scale in order to better view values around the resonant frequency. To obtain better accuracy, this simulation was set up to generate 10,000 data points over the plotting range. Using the Probe cursor, the resonant frequency f_0 is determined to be 5.035 KHz as predicted by PSpice. Shoulder frequency f_1 is found to be 4.953 KHz while f_2 is approximately 5.113 KHz.

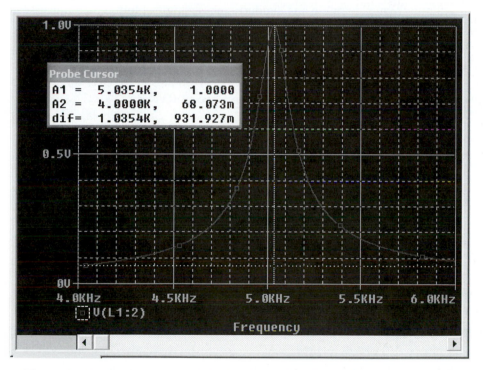

Figure 96 Simulation results of the bandpass filter focused on the pass band.

The resonant frequency may be obtained analytically for the circuit as $f_0 = \dfrac{1}{2\Pi\sqrt{LC}}$.

Plugging in the values for L and C, we obtain $f_0 = 5.033$ KHz which is slightly lower than the value calculated by PSpice. The difference between the calculated resonant frequency and the value produced by PSpice is approximately 2 Hz, which amounts to 0.04% of the expected frequency. This value is within the default accuracy parameters set for PSpice analysis. Using the Simulation Setting window Options tab, users may change the default accuracy settings and either increase or decrease the accuracy of the simulation results. However, we must be very cautious when changing these values since improper settings can cause several difficulties including incorrect results and extremely long simulation times.

7.2   Bode Plot of the Frequency Response

In the previous section we described how to plot the frequency response of a circuit on both a linear scale as well as a logarithmic scale. The linear scale was useful for narrowing in on responses at a specific frequency while the logarithmic scale allowed us to get a good view of the overall form of the response. We now examine the plotting of voltage magnitudes in decibels versus a logarithmic frequency range and similarly, phase angles plotted on a linear scale versus a logarithmic range of frequencies.

Example 17: Determine the frequency response for the circuit in Figure 97 and develop a Bode plot showing the output characteristics. Vary the load resistance over three values 2, 20 and 200 Ω. Determine the resonant frequency for each value of RVARY.

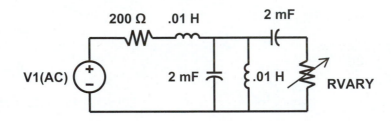

Figure 97 Example filter circuit with varying load resistance RVARY.

Solution 17: Entering the schematic into OrCAD Capture is done in much the same way as in earlier examples. Voltage source VAC is used as a steady-state sinusoidal source for frequency response analysis. Circuit elements are standard resistors, capacitors and inductors, each having a fixed value except for R2. The schematic is shown in Figure 98.

Load resistance R2 is set up to have a variable as the value for the resistance as opposed to a numerical value. Once the part R2 is in place, double-click on the value field of the resistor element in the schematic, and then enter the variable name {RVARY} in the Value field. The squiggly brackets are used to indicate that the parameter contained therein is to be evaluated as a variable parameter. To set up a mechanism to pass values to the variable parameter RVARY, we set up a global variable using a special schematic element called PARAM. To install the

PARAM part in the schematic, click on the Place Part button ⬚, and add the special.olb library. From the SPECIAL library, choose the part named PARAM and insert it into the schematic producing a part symbol titled <u>PARAMETERS:</u>. Double-click on the PARAMETERS symbol to open the Property Editor. We are now working to add the name RVARY to the list of global variables that are contained in the PARAM component. These variables can be placed in parts throughout the schematic and can be changed by simply modifying the value of the variable at only a single location. Click on the New Column... button to open the Add New Column dialog box. Type the name RVARY in the dialog box and enter the number 1 into the Value area providing a default value for RVARY. Click OK to return to the spreadsheet, which should look similar to the one in Figure 99. Now select the column labeled RVARY by clicking at the top of the column. Click on the Display button and fill out the dialog box so that the variable RVARY is displayed in the schematic as shown in the PARAMETERS component in Figure 98.

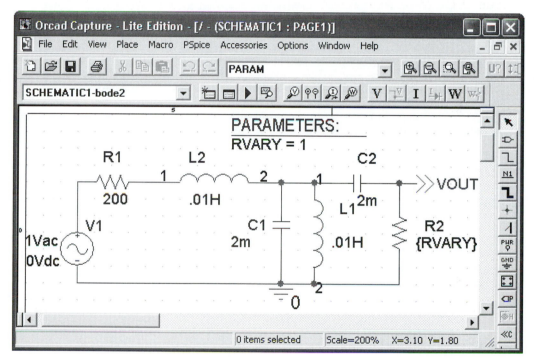

Figure 98 Schematic for the more complex filter.

We must next specify the analysis type and the range of frequencies over which we would like the analysis to take place. Open the simulation Settings window and set the analysis type to AC sweep and set the AC sweep type to be logarithmic plotted by decade. Then enter the start frequency to be 0.1 Hz, the end frequency to be 100 KHz and the points per decade to be 1000. The range of frequencies is chosen based on the size of the RLC elements and the expected low frequency behavior based on the size of the components.

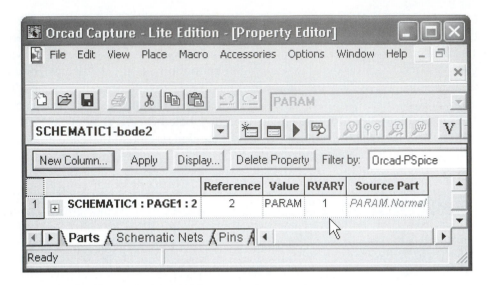

**Figure 99 Property editor window used to add the variable
RVARY to the PARAM component.**

Next, we set up the simulation so that the value of RVARY is changed to 2, 20, and 200 according to the specification of the problem. In effect what we are doing in this step is setting up three different simulations, each using the same frequency sweep range. Data from the three analyses is deposited into a common output file to be plotted by Probe. In the Simulation Settings window, click on the check box labeled Parametric Sweep in the Options area. This brings up a new area as shown in Figure 100 which is used to enter values related to sweeping the global variable RVARY previously set up in the PARAMETERS schematic component. Click on the radio button labeled Global Parameter and fill in the Parameter Name to be RVARY. Next, choose the Sweep Type to be Value List and fill in the values 2, 20, 200 in the list area as shown in Figure 100.

Click OK and then run the simulation. Once the simulation completes and the Probe window is displayed, a dialog box indicates that there are three sections to the output with all three sections highlighted. Click OK to indicate that Probe is to plot data from all three sections of the simulation. The Probe display now comes up empty and we must add traces to the window in order to view the output. From the menu click Trace>Add Trace…. At the bottom of the window, type in the expression db(V(VOUT)) displaying the frequency response of the output voltage VOUT in decibels for the full frequency sweep range. Each of the different values of RVARY produces a distinct plot at frequencies below the resonant frequency, but the curves essentially coincide at frequencies above the resonant frequency. We also notice that the response changes at the rate of approximately 35 decibels per decade of frequency on either side of the resonant frequency. We set the magnitude of the input V1 to be one, so that the curves readily represent the voltage magnitude transfer characteristics for the circuit. The Probe display should look similar to Figure 101.

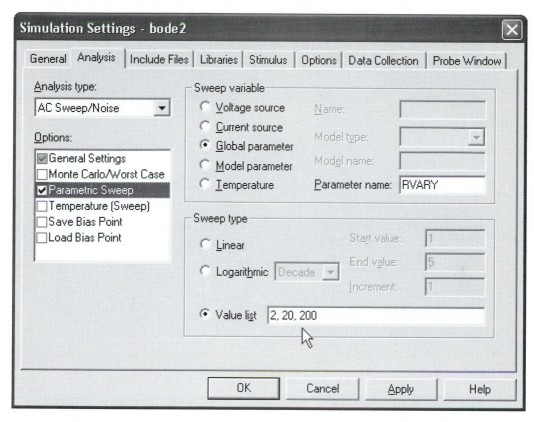

Figure 100 Parametric sweep settings used to change the value RVARY.

We are also interested in the phase of the output as it relates to the input and for this reason, the phase of the input V1 was set to zero. To plot the phase of the output on the same graph as the magnitude of VOUT, from the Probe menu, click on Plot>Add Plot to Window. Notice that the symbol SEL>> appears to the left of the empty axis to indicate that it is the currently selected coordinate system. Click on Trace>Add Trace and type in VP(VOUT) to plot the phase of the output as a function of frequency. The composite plot is shown in Figure 102 where we can observe the relationship between the changing phase angle and the magnitude of VOUT as a function of frequency. Notice the notation added to the composite plot to indicate which curve represents the response with R2 = 2. Menu items from Plot>Label are drawn on the plot for these annotations. Using the Probe cursor we find the resonant frequency to be approximately 27.3, 35.4, and 35.5 Hz for the load resistance 2, 20, and 200, respectively. The magnitude ranges from roughly 0 db to -120 db while the phase angles range from -180° to +180°.

In this chapter we described methods for using PSpice to produce the frequency response of a circuit when the applied input signal is a steady-state sinusoidally varying signal. This type of analysis is used in many different situations, especially when filter circuits are being designed.

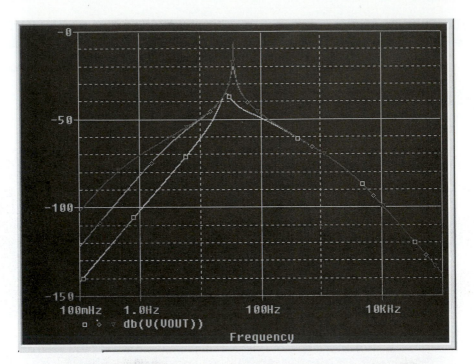

Figure 101 Plot of VOUT in decibels for the filter of Figure 98.

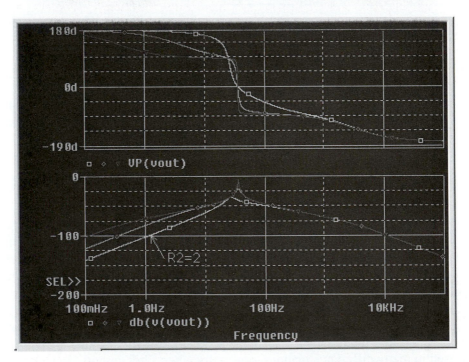

Figure 102 Bode plot of the magnitude and phase angle of VOUT.

Chapter 8 Fourier Series

Many circuits are designed to process non-sinusoidal signals. Although these signals are non-sinusoidal, they are still periodic, repeating certain patterns over a certain period T. This chapter is concerned with using PSpice to support the analysis of these types of signals. We will describe how to express non-sinusoidal signals in the terms of sinusoids which can then be analyzed using standard techniques such as phasor analysis.

8.1　Basic Analysis

Linear circuits that have a steady-state periodic signal applied to their inputs will produce an output that is periodic. As might be expected, the period of the output signal is equal to the period of the input signal. To break these periodic signals into their sinusoidal components, we use a Fourier series to describe each of the individual sinusoids, which when summed together produce the non-sinusoidal periodic signal. The Fourier series for a function $f(t)$ can be expressed as:

$$f(t) = a_0 + \sum_{n=1}^{\infty} a_n \sin(n\omega_0 t + \theta_n)$$

where ω_0 is the fundamental frequency in radians per second and is related to the period of the signal by

$$\omega_0 = \frac{2\Pi}{T} = 2\Pi f$$

where T is the period and f is the frequency of the periodic signal.

The task at hand is to determine the values of the individual coefficients a_1, a_2, a_3 ,…, θ_1, θ_2, θ_3,… corresponding to the specific $f(t)$.

An example of a typical non-sinusoidal waveform is shown in Figure 103. The waveform is periodic with a value of $T = 2$ ($\omega_0 = \pi$) and a Fourier series description of:

$$f(t) = \frac{1}{2} + \frac{2}{\Pi} \sum_{k=1}^{\infty} \frac{1}{n} \sin n\Pi t, \quad \text{where } n = 2k\text{-}1$$

or

$$f(t) = \frac{1}{2} + \frac{2}{\Pi} \sin \Pi t + \frac{2}{3\Pi} \sin 3\Pi t + \frac{2}{5\Pi} \sin 5\Pi t + \dots$$

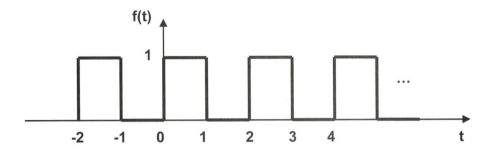

Figure 103 Example waveform for Fourier series analysis.

The equation above shows only the constant component (dc component), and the first three time-dependent components. Notice that the coefficient a_i becomes very small after the third series term and does not influence the shape of the resulting function $f(t)$ very much.

8.2   *Fourier Circuit Analysis*

OrCAD PSpice can calculate the Fourier series coefficients for a waveform in two ways: the PSpice program can perform a discrete Fourier transform (DFT) on a waveform as part of its transient analysis calculations, or the Probe plotting package can perform a fast Fourier transform (FFT) on the data resulting from a transient analysis. The next example describes how these operations can be performed.

Example 18: Use the waveform shown in Figure 104 as input to the circuit in Figure 105. Perform a transient analysis on the circuit and then determine the Fourier series expression for both the input and output waveforms.

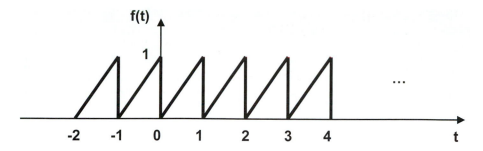

Figure 104 Sawtooth waveform with T = 1.

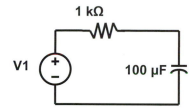

Figure 105 Determine the Fourier series expression for the response of this circuit.

Solution 18: We develop a schematic for the circuit much like earlier schematics using the VPULSE voltage source to produce the input voltage waveform shown in Figure 104. In order to produce the sawtooth waveform, VPULSE parameters must be set as follows: V1 = 0, V2 = 1, TD = 0. Setting the rise time TR = 1 provides the sloped portion of the waveform while setting the fall time TF = 0 provides the vertical portion of the sawtooth. Fixing the pulse width PW = 1 and period PER = 1 produces a waveform that approximates the ideal sawtooth waveform. The schematic in Figure 106 shows the circuit with these parameters set.

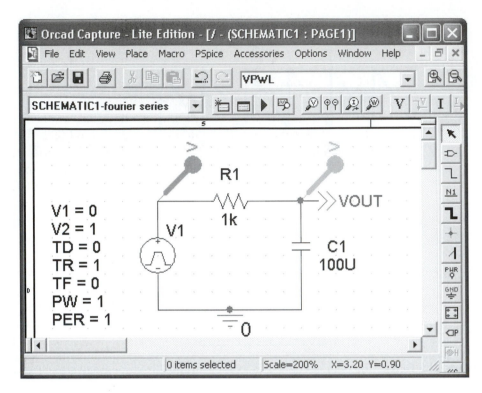

Figure 106 Schematic with VPULSE to produce a sawtooth voltage input.

The next step is to set up the Simulations Settings window for the appropriate transient analysis followed by the calculation of the Fourier series coefficients. We open the Simulation Settings window and choose the analysis type to be Transient Analysis. We set the value of TSTOP to be five seconds so that the circuit response has stabilized in its steady-state—i.e., all effects due to the given initial conditions $V_c = 0$ have been eliminated and the only thing affecting the circuit is the sawtooth input. (To further illustrate this point, try running the simulation by setting an initial condition of $V_c = 10$ and you will see that the effect of the initial condition on C1 is lost after about two cycles of the input waveform). PSpice calculates the Fourier coefficients on the output waveform starting at time TSTOP and working backward one full cycle of the fundamental frequency. It is necessary to ensure that the waveform has reached steady-state in order to perform the Fourier analysis properly.

Time TSTART, or the *start saving data after time,* is set to zero in order to observe all of the waveform. The Maximum Step Size parameter sets the internal step size (variable *istep*) for the numerical routines that PSpice uses to solve the circuit. This value will be determined automatically if the field is left blank. However, for this problem we must specify a step size and we choose 0.001. The reason for this choice is that PSpice restricts the minimum size for either TR or TF to be no less than the internal time step called *istep*. If we specify a value for TR or TF that is less than *istep*, the value of *istep* is used in its place. Therefore, in order to insure that we have a very fast fall time in our sawtooth input, we specify a value for *istep* rather than using the one PSpice would calculate internally.

Now that we have the transient analysis set up, we must next set up the Fourier analysis. Click on the button labeled Output File Options… and set the system to print values in the output file every 0.01 seconds. Check the box labeled Perform Fourier Analysis and specify the center frequency to be 1 Hz, which is the fundamental frequency for our input waveform. Set the number of harmonic coefficients to be calculated to 10 in order to get a good feel for the pattern of the series of coefficients. Finally, specify the Output Variables for which Fourier coefficients will be calculated; in this case we choose the voltage across the capacitor and the input voltage. The Transient Output File Options window should look similar to the one in Figure 107.

PSpice generates the simulation file shown in Figure 108 in response to the schematic and simulation settings input provided for this problem. Notice the statement that specifies the transient analysis (.TRAN) includes parameters taken directly from the Simulation Settings window along with the *istep* parameter set in the Transient Output File Options window. Similarly, the .FOUR statement is included in the simulation file, and gains its parameters, based on information provided in the Transient Output File Options window. The statement that describes the circuit element V1 includes the specifications for the transient behavior set in the properties dialog box for the element V1. One can easily conclude that one of the functions of the capture software is to assemble all of the relevant information related to the topology of the circuit as well as collecting information related to the simulation parameters used by PSpice.

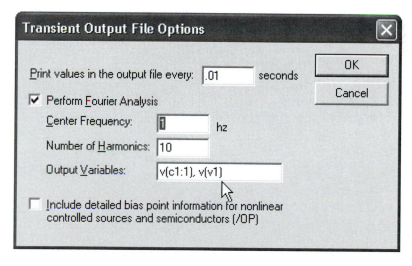

Figure 107 Dialog box used to set up a Fourier analysis.

```
*Analysis directives:
.TRAN .01 5 0 .001
.FOUR 1 10 v([VOUT]) v(V_V1)
.PROBE V(*) I(*) W(*) D(*) NOISE(*)
.INC ".\fourier series-SCHEMATIC1.net"
**** INCLUDING "fourier series-SCHEMATIC1.net"
****
* source FOURIER SERIES
C_C1        VOUT 0  100U
R_R1        N00176 VOUT  1k
V_V1        N00176 0
+PULSE 0 1 0 1 0 1 1
```

Figure 108 Simulation file for Fourier analysis example.

After the simulation is run, Probe displays a window similar to the one in Figure 109. Waveforms representing the voltage across the capacitor and the voltage produced by input source V1 are included in the plot because of the voltage probes placed in the original schematic. The input waveform is the sawtooth that is expected based on the parameters supplied for V1. The second waveform shows that the output waveform does not fall all the way to zero before the input switches upward and begins to drive the output back toward the value one. The lag is due to the size of the RC charging time constant. Further examination of the output waveform indicates that the output voltage reaches steady-state operation after roughly one period as evidenced by the fact that the output waveform is nearly identical for each subsequent input period.

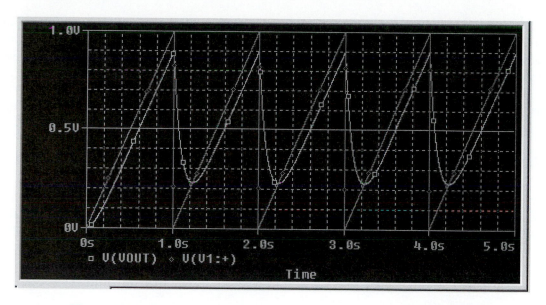

Figure 109 Transient output showing both input and output waveforms.

Fourier coefficients for the two variables selected in the Transient Output File Options window are shown in the printed output file which can viewed by clicking the View Simulation Output File button ⬚ . Scrolling down in the output file you will find the coefficients for the input signal V1 as shown in Figure 110. Using the first three harmonic coefficients along with the dc component, we can construct the Fourier series equation that describes the input waveform as:

$$f(t) = 0.495 + 0.318\sin(\Pi t - 178°) + 0.159\sin(3\Pi t - 176°) + 0.106\sin(5\Pi t - 175°) + ...$$

The ideal sawtooth waveform shown in Figure 104 can be represented by the Fourier series expansion:

$$f(t) = \frac{1}{2} - \frac{1}{\Pi}\sum_{k=1}^{\infty}\frac{1}{n}\sin 2\Pi nt$$

Examining these two expressions we see that the major difference appears to occur in the phase of the sinusoid. Coefficients generated by PSpice are positive, whereas the theoretical expression shows a negative sign in front of the coefficients. However, when you include consideration of the phase angle in the PSpice generated coefficients, the expression provides a very good approximation to the theoretical form. This is due to the fact that $\sin(x-180°) = -\sin(t)$. Phase angles for the first three coefficients are roughly 180° and thus produce the effect of negating the associated term. The difference between the first three phase angles of -178°, -176° and -175° and the expected value of -180° is likely due to the inaccuracy in describing the sawtooth waveform.

Remember, because the minimum fall time TF is *istep*, the falling edge of the sawtooth does not have an infinite slope as would be expected in the original theoretical calculation. Reduced numerical accuracy also accounts for some of the reduced accuracy of the calculated phase angles.

Fourier series coefficients for the voltage output VOUT are shown in Figure 111. The Fourier series expression for the output voltage is

$$VOUT(t) = 0.4998 + 0.270\sin(\Pi t - 148°) + 0.099\sin(3\Pi t + 129°) + 0.050\sin(5\Pi t + 118°) + ...$$

The reader should confirm that the waveforms described in the expressions above for the input and output voltages match the transient analysis output by plotting the expression for the Fourier series expansion using a program like Matlab.

```
DC COMPONENT =    4.950487E-01

HARMONIC    FREQUENCY     FOURIER     NORMALIZED     PHASE        NORMALIZED
   NO         (HZ)       COMPONENT    COMPONENT      (DEG)       PHASE  (DEG)

    1       1.000E+00    3.184E-01    1.000E+00    -1.782E+02     0.000E+00
    2       2.000E+00    1.593E-01    5.002E-01    -1.764E+02     1.800E+02
    3       3.000E+00    1.063E-01    3.338E-01    -1.747E+02     3.600E+02
    4       4.000E+00    7.978E-02    2.506E-01    -1.729E+02     5.400E+02
    5       5.000E+00    6.392E-02    2.008E-01    -1.711E+02     7.200E+02
    6       6.000E+00    5.336E-02    1.676E-01    -1.693E+02     9.000E+02
    7       7.000E+00    4.583E-02    1.440E-01    -1.675E+02     1.080E+03
    8       8.000E+00    4.020E-02    1.263E-01    -1.657E+02     1.260E+03
    9       9.000E+00    3.584E-02    1.126E-01    -1.640E+02     1.440E+03
   10       1.000E+01    3.235E-02    1.016E-01    -1.622E+02     1.620E+03
```

Figure 110 Fourier series coefficients for the sawtooth input waveform.

```
DC COMPONENT =    4.998199E-01

HARMONIC    FREQUENCY     FOURIER    NORMALIZED     PHASE       NORMALIZED
   NO          (HZ)      COMPONENT   COMPONENT      (DEG)      PHASE (DEG)

    1       1.000E+00    2.696E-01   1.000E+00    1.479E+02    0.000E+00
    2       2.000E+00    9.921E-02   3.680E-01    1.286E+02   -1.672E+02
    3       3.000E+00    4.985E-02   1.849E-01    1.181E+02   -3.256E+02
    4       4.000E+00    2.955E-02   1.096E-01    1.120E+02   -4.797E+02
    5       5.000E+00    1.944E-02   7.211E-02    1.080E+02   -6.316E+02
    6       6.000E+00    1.374E-02   5.095E-02    1.052E+02   -7.823E+02
    7       7.000E+00    1.022E-02   3.790E-02    1.032E+02   -9.322E+02
    8       8.000E+00    7.901E-03   2.930E-02    1.017E+02   -1.082E+03
    9       9.000E+00    6.297E-03   2.336E-02    1.006E+02   -1.231E+03
   10       1.000E+01    5.143E-03   1.908E-02    9.963E+01   -1.380E+03
```

Figure 111 Fourier series coefficients for the output voltage of the RC circuit.

As a further example of generating the Fourier coefficients, we perform the same simulation as above, but with the size of the capacitor increased to 500 µF. Transient analysis output for this configuration is shown in Figure 112. Output voltage VOUT dramatically lags behind the input voltage because of the increased size of the capacitor. Fourier coefficients for the new output waveform are shown in Figure 113. The Fourier series expression for the output voltage is

$$VOUT(t) = 0.4997 + 0.097\sin(\Pi t - 108°) + 0.025\sin(3\Pi t + 99°) + 0.011\sin(5\Pi t + 96°) + ...$$

Notice that the frequency components that cause the output waveform to oscillate in the neighborhood of ½ volt are spread out fairly well across the first 10 harmonics.

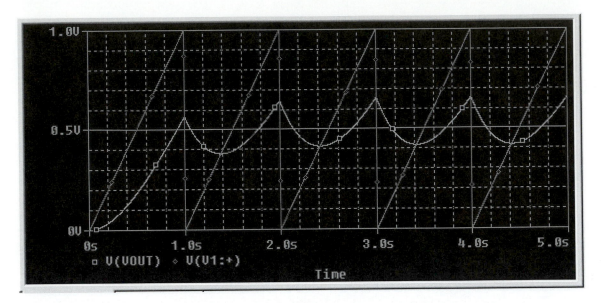

Figure 112 Transient output with C1 = 500 μF.

DC COMPONENT =   4.996685E-01

HARMONIC NO	FREQUENCY (HZ)	FOURIER COMPONENT	NORMALIZED COMPONENT	PHASE (DEG)	NORMALIZED PHASE (DEG)
1	1.000E+00	9.657E-02	1.000E+00	1.078E+02	0.000E+00
2	2.000E+00	2.504E-02	2.593E-01	9.928E+01	-1.163E+02
3	3.000E+00	1.122E-02	1.162E-01	9.641E+01	-2.269E+02
4	4.000E+00	6.339E-03	6.564E-02	9.502E+01	-3.361E+02
5	5.000E+00	4.071E-03	4.215E-02	9.423E+01	-4.446E+02
6	6.000E+00	2.837E-03	2.937E-02	9.374E+01	-5.529E+02
7	7.000E+00	2.092E-03	2.166E-02	9.342E+01	-6.610E+02
8	8.000E+00	1.608E-03	1.666E-02	9.319E+01	-7.690E+02
9	9.000E+00	1.276E-03	1.322E-02	9.303E+01	-8.770E+02
10	1.000E+01	1.039E-03	1.076E-02	9.294E+01	-9.848E+02

Figure 113 Fourier series coefficients for the output waveform with C1 = 500 μF.

Chapter 9 Conclusion

9.1   Common Mistakes

PSpice is a very useful tool for supporting circuit analysis. However, it must be applied carefully and the user must always be aware of the reasonableness of the results. Subtle mistakes can cause incorrect results. All output should be evaluated to determine if it is in the expected range of solutions. A few of the typical mistakes that are made in setting up a simulation are noted below in the hopes that the reader will avoid these common errors.

- Leaving out the Ground symbol is the number one mistake. It is easily remedied by inserting the symbol and re-running the simulation.
- Typos in the source file: wrong node numbers, wrong units (e.g., the value of a capacitor of 43 picofarads, incorrectly specified as 43, instead of 43p.)
- Typing the <u>letter</u> O instead of the <u>number</u> 0 (zero as in 10)
- Confusing M for mega instead of MEG. A 6 megaohm resistor should be specified as 6MEG and not 6M. (M or m stands for milli.)
- Writing 1F to represent one farad of capacitance. In fact, the F represents the abbreviation femto resulting in 1×10^{-15} units of capacitance.
- Omitting the final carriage return after the .END statement. This applies to those who enter their netlist information as a text file rather than using the schematic capture software.

9.2   Summary

Using powerful CAD tools is an integral part of modern engineering practice. Several different software packages designed to simulate the behavior of electrical circuits exist. SPICE distinguishes itself by being the industry standard for those designing integrated circuits. OrCAD PSpice is one of the most popular derivatives of the original version of SPICE. Our purpose in introducing the PSpice software to those who are beginning their studies at the linear circuit level is to allow them to become comfortable with many of the various aspects of this powerful problem solving tool. Readers should have received an understanding of the fundamentals of the structure of PSpice-supported analysis along with the various ways in which the tool can be integrated into an analysis and synthesis methodology.

Bibliography

- PSpice Reference Guide, Release 9.2, Copyright Cadence Design Systems, Inc., 1985-2000.
- The Spice Book, A. Vladimirescu, John Wiley & Sons, New York, NY, 1994.
- Semiconductor Device Modeling with Spice, 2nd Ed., G. Massobrio and P. Antognetti, McGraw-Hill, NY, 1993.
- Inside SPICE, Overcoming the Obstacles of Circuit Simulation, R.M. Kielkowski, McGraw-Hill, Inc., New York, 1994.
- SPICE, Practical Device Modeling, R.W. Kielkowski, McGraw-Hill, Inc., New York, 1995.
- "The Simulation of MOS integrated circuits using SPICE2," A. Vladimirescu, and S. Liu, University of California, Berkeley, ERL Memo UCB/ERL M80/7, March 1981.
- "Simulation Program with Integrated Circuit Emphasis (SPICE)," L.W. Nagel, D.O. Pederson, University of California, Berkeley, ERL Report No. ERL M383, April 1973.

Appendix: Files Used by PSpice

The list below contains the names and file types for many of the files used by PSpice. This information may be of use in organizing your files or troubleshooting difficulties encountered when using the OrCAD system.

- ***.opj**: the project file which contains information about all of the design entities associated with a specific project
- ***.dsn**: contains information about specific design entities that make up a project
- ***.lib** or ***.olb**: library files containing schematic parts, component models or user defined entities used in the capture and simulation process
- ***.cir**: PSpice netlist file
- ***.net**: netlist file acceptable by many CAD packages; must have further information appended to be ready for PSpice simulation
- ***.sch:** file containing information that describes the circuit schematic
- ***.out**: contains the text output generated by PSpice
- ***.dat**: contains the numerical data output by PSpice and plotted by Probe
- **analog.slb**: contains resistors (R), capacitors (C), inductors (L), dependent sources (E, G, F and H)
- **source.slb**: use for independent sources, Ex. VSRC (Simple voltage source)
- **port.slb**: contains elements such as ground: Ex. GND_ANALOG; this is used as the reference node (node 0) and must be added to the schematic

Index